# Independent CHST Exam Prep 2027-2028:

## Professional Study Guide with 400 Practice Questions, Detailed Explanations for Construction Health and Safety Technician Certification Success

**Yolanda Mensah**

## Author Introduction

In construction environments where conditions shift rapidly and risks are ever-present, safety expertise must be both precise and practical. The **Construction Health and Safety Technician (CHST) Exam** serves as a respected benchmark for professionals committed to maintaining safe, compliant, and well-managed job sites. This study guide was developed to meet the demands of that benchmark—equipping candidates with the knowledge, applied understanding, and confidence required to succeed both on the exam and in the field.

The CHST credential assesses a candidate's ability to recognize hazards, interpret and apply safety regulations, conduct job hazards analyses, support safety training initiatives, and contribute meaningfully to incident prevention efforts. It is not simply a test of memorization; it evaluates judgment, problem-solving ability, and practical understanding of construction safety operations. This guide mirrors the structure and intent of the exam blueprint, organizing content into clearly defined domains that reflect real-world responsibilities. Each chapter is designed to translate regulatory language and technical standards into actionable insights that candidates can readily apply in exam scenarios.

The author of this guide is an experienced construction safety professional who has spent years working within active job sites, collaborating with project managers, field supervisors, and compliance teams to strengthen safety systems. This firsthand experience shapes the approach of the material—prioritizing clarity, practicality, and relevance. Rather than presenting abstract concepts, the author integrates industry-based examples and applied reasoning strategies to help candidates understand not only what the standards require, but why they matter in daily operations.

In developing this resource, the author worked in collaboration with a dedicated team of safety educators, subject-matter specialists, and instructional designers. Together, they analyzed examination objectives, reviewed regulatory updates, and incorporated proven instructional methods to ensure the content remains accurate, focused, and aligned with current industry expectations. This collaborative effort ensures that candidates receive preparation grounded in both technical precision and educational effectiveness.

At the core of this guide is the author's commitment to advancing safety leadership within the construction industry. The CHST credential represents professional accountability and dedication to worker protection. Through structured explanations, targeted practice strategies, and domain-specific insights, the author aims to support candidates not only in passing the exam, but in strengthening their role as proactive contributors to safer construction environments.

# Table of Content

This study guide is an independent educational resource developed to assist candidates preparing for the Construction Health and Safety Technician (CHST) examination. It is not affiliated with, endorsed by, sponsored by, or approved by the Board of Certified Safety Professionals (BCSP) or any other certifying organization. All trademarks are the property of their respective owners.

Every effort has been made to ensure that the information contained in this publication is accurate, current, and aligned with the CHST examination content outline at the time of publication. However, standards, regulations, and examination objectives may change over time. The publisher and authors make no representations or warranties regarding the completeness or accuracy of the information provided and disclaim any liability for errors, omissions, or outcomes resulting from the use of this material.

This guide is intended solely as a study aid to support exam preparation. It is not a substitute for official regulatory standards, professional training, or practical field experience. Readers are encouraged to consult current OSHA regulations, applicable safety standards, and official CHST examination resources when preparing for certification.

The strategies, practice questions, and review materials contained herein are designed to enhance understanding and reinforce key concepts relevant to the CHST examination domains. Success on the exam depends on individual preparation, professional experience, and mastery of the subject matter.

## Disclaimer

CHST® is a registered trademark of the Board of Certified Safety Professionals, Inc. This publication is an independent study tool and is not affiliated with or endorsed by BCSP®

# Chapter 1: Hazard and Risk Identification and Control

## Understanding the Foundations of Hazard Recognition

Hazard recognition lies at the heart of construction safety management. Every task, environment, and material presents potential risks that must be identified before they cause harm. The ability to detect hazards depends on a disciplined process of observation, analysis, and communication. A hazard is any condition or practice that has the potential to cause injury, illness, or property damage. Recognition begins with knowing where to look—examining the work environment, reviewing processes, and engaging with workers who understand their tasks intimately. Effective hazard identification is not achieved by checklists alone; it relies on critical thinking and awareness of how conditions change as work progresses. Construction sites are dynamic, with multiple trades, weather conditions, and materials interacting simultaneously. This constant flux increases exposure potential and demands continuous reassessment. A competent safety professional observes work flow, identifies both obvious and latent hazards, and classifies them by likelihood and severity. The recognition process extends beyond the visible to include ergonomic stresses, chemical exposures, and psychosocial risks that influence decision-making. Every safety system begins with accurate recognition, for without it, control measures become reactionary rather than preventive. Understanding the sources, patterns, and consequences of hazards transforms identification from a routine inspection into a strategic discipline.

## The Nature of Risk and Its Relationship to Control

Risk represents the probability and consequence of harm arising from a hazard. While a hazard is an inherent condition, risk depends on exposure and control. Construction environments amplify risk because human behavior, equipment condition, and environmental change continuously affect exposure levels. Assessing risk requires both qualitative judgment and quantitative evaluation. Techniques such as risk matrices, scoring systems, or hierarchy-based prioritization help determine which hazards demand immediate attention. The relationship between risk and control is direct: effective controls reduce either the likelihood or severity of harm, ideally both. However, control is not static; it must adapt as tasks, personnel, or materials change. Over time, even well-designed systems degrade if not reviewed and maintained. Risk management is therefore cyclical—identify, assess, control, monitor, and review. Understanding this relationship enables safety technicians to apply proportional responses, ensuring resources target the most significant threats. The goal is not the elimination of all risk, which is impossible, but the achievement of tolerable levels that allow safe productivity. Every risk decision should balance operational need with protection of life and health, recognizing that uncontrolled risk equates to unmanaged liability.

## The Hierarchy of Controls as a Strategic Framework

The hierarchy of controls provides a structured approach for mitigating hazards by prioritizing interventions according to their effectiveness. At its top stands elimination—removing the hazard entirely from the process. Substitution follows, replacing hazardous materials or methods with safer alternatives. Engineering controls isolate people from hazards through design features such as barriers, ventilation, or automated systems. Administrative controls adjust how people work, introducing training, scheduling, or procedural safeguards. Personal protective equipment forms the final barrier when no other control is feasible. The framework is both philosophical and practical: it reinforces that protective gear should never compensate for poor design or planning. In construction, applying this hierarchy requires collaboration between project managers, supervisors, and workers during pre-job planning. A well-implemented control hierarchy also considers human factors such as fatigue, experience level, and situational awareness. Its strength lies in prevention rather than reaction. Technicians who internalize this model evaluate every job task by asking which controls can eliminate exposure before resorting to equipment or rules. Overreliance on procedural or personal protection often signals missed opportunities for higher-level interventions. True mastery of control hierarchy transforms safety from compliance to design.

## Principles of Job Hazard Analysis

Job Hazard Analysis (JHA), sometimes called Job Safety Analysis, is the systematic examination of individual work tasks to identify hazards and define safe operating procedures. It bridges planning and performance by converting general risk awareness into task-specific action steps. A JHA begins by breaking a job into sequential steps, observing how workers interact with tools, materials, and the environment. For each step, potential hazards are identified—mechanical, chemical, ergonomic, or environmental—and corresponding control measures are assigned. The process encourages collaboration between those performing the work and safety professionals, ensuring that procedures reflect real-world conditions rather than theoretical assumptions. A well-developed JHA evolves with the job; it is reviewed whenever equipment, personnel, or conditions change. The technique also serves as a training document, reinforcing safe habits and standardizing expectations across crews. Beyond compliance, a JHA cultivates ownership of safety by involving workers directly in hazard recognition. This engagement not only improves accuracy but also strengthens safety culture, as individuals learn to anticipate risk before it materializes. The JHA embodies proactive safety—predicting incidents before they occur through methodical task analysis and communication.

## Common Physical Hazards and Their Controls

Construction sites host a range of physical hazards—each capable of producing severe injury if uncontrolled. These include falls from height, struck-by and caught-in incidents, electrical exposures, vibration, and excessive noise. Controlling such hazards begins with understanding how they arise. Falls typically result from poor edge protection, unstable scaffolding, or improper harness use. Struck-by events stem from moving vehicles, suspended loads, or unguarded equipment. Effective control measures integrate design and behavioral interventions: guardrails, toeboards, exclusion zones, lockout-tagout systems, and tool tethering all reduce likelihood of harm. Noise and vibration require technical evaluation through monitoring devices and engineering solutions such as dampeners or sound barriers. The principle of safety through design emphasizes prevention during planning stages, where hazards can be engineered out rather than managed in the field. A safety professional must also recognize interaction hazards—situations where multiple trades overlap and one group's task endangers another. Clear communication and sequencing of work prevent these compounded risks. Physical hazard control reflects the essential balance of foresight, supervision, and continuous verification. Each preventive measure, when applied thoughtfully, reinforces a network of protection that allows safe and productive construction.

## Managing Chemical and Environmental Exposures

Chemical and environmental hazards pose unique challenges on construction sites. Substances such as solvents, coatings, adhesives, and dusts introduce inhalation and contact risks, while environmental extremes—heat, cold, or radiation—affect worker endurance and health. Managing these exposures begins with identification through labels, safety data sheets, and the Globally Harmonized System of

Classification and Labeling of Chemicals (GHS). Once recognized, exposure assessment follows using air monitoring, sampling, or comparison against occupational exposure limits. Control relies first on substitution or containment—selecting low-toxicity materials, enclosing mixing processes, or using local exhaust ventilation. Administrative measures, such as rotation schedules or restricted access, further limit duration and intensity of exposure. Where residual risk persists, personal protective equipment such as respirators, gloves, or protective clothing becomes essential. Environmental conditions demand similar layered protection: acclimatization programs, hydration planning, and rest intervals for heat; insulated clothing and wind protection for cold. Technicians must also monitor weather forecasts and site microclimates, adjusting work methods accordingly. Proper storage, labeling, and disposal of materials prevent secondary contamination. The management of these hazards reflects professional discipline—translating laboratory data into field-ready safeguards that preserve worker health.

## Ergonomics and Human Factors in Construction Safety

Ergonomics, often associated with office environments, is equally vital in construction where repetitive motions, awkward postures, and excessive loads can cause chronic injury. Human factors research extends this concept by examining how physical and cognitive capacities interact with tools, work design, and environmental pressures. Applying ergonomic principles begins with evaluating body mechanics in lifting, carrying, or operating machinery. The objective is to design tasks that fit human capability rather than forcing adaptation. Techniques include mechanical aids, adjustable platforms, and proper sequencing to reduce strain. Beyond the physical, cognitive ergonomics addresses decision fatigue, attention limits, and perception of risk. Fatigue and distraction contribute to a significant percentage of construction incidents. Therefore, scheduling, lighting, and workload management become ergonomic interventions as well. By aligning task design with natural human function, safety technicians not only prevent musculoskeletal injuries but also improve efficiency and morale. Recognizing that workers are the central component of every system reframes safety from control of behavior to design for capability. Ergonomics transforms the site into a workspace that supports human performance instead of challenging it.

## The Role of Safety Observation and Monitoring

Observation and monitoring are continuous processes that validate whether control measures remain effective. Even the best-designed systems degrade under production pressure, equipment wear, or complacency. Regular observation ensures early detection of emerging risks. Effective safety observation is not policing—it is coaching. The observer's role is to recognize safe behavior, identify at-risk actions, and provide immediate constructive feedback. Monitoring also involves the use of instruments and technology—noise dosimeters, gas detectors, or vibration meters—to measure conditions objectively. These data form the basis for trend analysis, identifying where additional control or training may be required. Transparent reporting of findings builds trust and accountability. Workers who understand that observations are intended to protect, not punish, participate openly, enhancing accuracy of results. Over time, consistent monitoring cultivates situational awareness across all levels of the organization. The ultimate goal is self-regulation, where every worker internalizes observation habits and responds to risk instinctively. Through active monitoring, safety remains dynamic and resilient, adjusting to the evolving complexity of construction projects.

## Integrating Safety through Design and Planning

Safety through design, sometimes called prevention through design, integrates hazard elimination into the earliest stages of project planning. By addressing safety before construction begins, many risks are removed at their source. Design professionals collaborate with safety technicians to evaluate construction methods, materials, and sequences for potential hazards. For instance, prefabrication reduces work at height; modular components minimize manual handling; built-in anchorage points simplify fall protection. The approach recognizes that design decisions influence downstream safety outcomes more profoundly than any rule enforced later in the process. Integrating safety into design requires systematic review meetings, constructability analyses, and continuous feedback between engineers and field supervisors. Documented design risk assessments become part of project records, ensuring transparency and accountability. This proactive strategy not only prevents injuries but also enhances efficiency by reducing rework and downtime. Safety through design transforms safety from an operational constraint into a design criterion equal in importance to cost, quality, and schedule.

## The Culture of Shared Responsibility

The success of hazard control depends on a culture where every individual accepts responsibility for safety outcomes. Shared responsibility means that management provides resources and leadership, supervisors enforce procedures and model behavior, and workers actively participate in identifying and resolving hazards. Culture develops through consistent actions, open communication, and visible commitment. A mature safety culture encourages reporting of near misses without fear of blame, recognizing that such events provide valuable learning opportunities. Leadership reinforces this mindset by acknowledging safe performance and addressing violations fairly and promptly. Over time, collective behavior shifts from compliance to commitment. When workers correct hazards without prompting or remind peers to use protective equipment, culture has taken root. The safety technician's role is to nurture this environment through engagement, respect, and persistence. Culture cannot be mandated; it must be cultivated. Once achieved, it becomes the most powerful control of all—an invisible network of shared vigilance that sustains safety long after formal inspections end.

## Continuous Improvement and the Feedback Loop

Hazard and risk control are not static accomplishments but evolving practices. Continuous improvement ensures that lessons from incidents, audits, and observations feed back into future planning. The process begins with data collection—incident reports, inspection results, and worker feedback—followed by analysis to identify patterns and root causes. Corrective actions are implemented, monitored for effectiveness, and adjusted as necessary. This loop transforms safety from a compliance-driven activity into a learning system. It also aligns with organizational quality and environmental management frameworks, allowing integration across departments. Effective feedback depends on accurate documentation and transparent communication. Technicians must ensure that findings are not buried in reports but translated into real operational changes. The goal is perpetual advancement: each project safer than the one before. Continuous improvement reflects humility within safety practice—the recognition that perfection is unattainable, but progress is always possible. Through this mindset, hazard control evolves alongside technology, experience, and human understanding, sustaining the integrity of the CHST professional standard.

# Chapter 2: Emergency Preparedness, Incident Investigation, and Response

## Building the Foundation of Emergency Preparedness

Emergency preparedness is a structured commitment to anticipate, plan, and mitigate incidents before they escalate into disasters. In construction safety management, it represents the bridge between prevention and response. An emergency can emerge from multiple triggers—fire, collapse, environmental contamination, equipment failure, or medical events. Preparedness begins with recognition of site-specific hazards and extends to development of response systems proportionate to risk. A robust emergency framework identifies who will act, what they will do, and how communication will flow under stress. Preparedness is never a static document but a living process that evolves with project scope, seasonal changes, and workforce composition. Training is its cornerstone: without consistent drills, even the best-designed plan collapses under pressure. Preparedness also depends on coordination with external agencies such as fire departments and medical responders to ensure seamless interoperability. Ultimately, preparedness transforms anxiety into readiness—replacing uncertainty with practiced response. In a well-prepared workplace, emergencies do not expose disorganization; they reveal resilience, discipline, and professional foresight.

## The Architecture of the Emergency Action Plan

An Emergency Action Plan (EAP) translates risk awareness into structured command. It delineates responsibilities, communication channels, evacuation routes, and post-event procedures in written form accessible to all personnel. A comprehensive EAP reflects regulatory guidance and site realities, incorporating elements from OSHA, NFPA, and local emergency codes. Its architecture combines clarity and flexibility: clarity ensures that everyone understands their role; flexibility allows adaptation to unforeseen variables. Each plan designates incident command authority, assigns alternates, and defines triggers for escalation or external notification. Visual aids—maps, muster points, and alarm procedures—enhance usability. The process of writing the EAP is as valuable as the document itself because it compels teams to analyze vulnerabilities systematically. When reviewed through drills and debriefings, weaknesses become visible, prompting refinement. Effective plans also address recovery, ensuring continuity of operations once the immediate crisis passes. The EAP is therefore not a formality but a survival blueprint. It translates policy into coordinated human behavior under the most demanding conditions imaginable.

## Communication Systems and Coordination Under Pressure

In an emergency, communication becomes the lifeline that determines whether response efforts succeed or fail. Clear, concise, and verified information flow prevents confusion and duplication of effort. The challenge lies in the stress of the moment—alarms sounding, multiple responders acting, and evolving hazards. To maintain order, communication protocols must be predefined and practiced. Incident command systems establish hierarchy, ensuring messages flow through reliable channels. Modern construction sites increasingly employ radio networks, mass notification systems, and digital dashboards to transmit alerts. However, technology cannot replace discipline; every worker must know who to contact and what information to convey. Communication during emergencies also extends beyond the site boundary. Coordination with fire, medical, and environmental authorities ensures that response scales appropriately. Post-event communication with families, clients, and regulators further demonstrates accountability and transparency. Consistent terminology, verified reporting, and redundancy safeguard against misinformation. The art of emergency communication is not speed alone but precision—the ability to transmit the right message to the right person at the right time.

## Fire Protection and Prevention as Core Preparedness

Fire remains one of the most destructive and unpredictable emergencies on construction sites. Its prevention relies on disciplined control of ignition sources, fuel materials, and oxygen flow—the elements of the fire triangle. Hot work operations, electrical systems, and temporary heating devices represent common ignition points that demand strict permitting and supervision. Preventive measures such as segregation of flammables, regular inspection of fire extinguishers, and availability of water supplies are fundamental. Fire protection is not limited to equipment but extends to worker competence. Every individual must know how to identify fire risks, activate alarms, and operate extinguishers safely. Evacuation routes should be illuminated, unobstructed, and verified daily. Construction projects evolve rapidly, and fire plans must evolve with them. The introduction of new materials, such as insulation or coatings, changes combustion potential. Training, inspection, and communication maintain vigilance. Fire prevention exemplifies proactive safety—the recognition that controlling energy before ignition is far easier than battling flames after they start.

## Medical Emergencies and First Response

A construction site is a complex, high-energy environment where medical emergencies can arise from falls, equipment malfunctions, or sudden illness. The first few minutes after an incident are decisive in determining outcomes. A competent emergency medical plan ensures immediate aid is available until professional responders arrive. First-aid stations, stocked and accessible, are strategically located near high-risk zones. Designated first-aiders must be trained and refreshed regularly in accordance with occupational regulations. However, medical response extends beyond basic first aid—it includes rapid assessment, communication of critical information, and management of emotional stress among witnesses. A clear chain of responsibility ensures that injured workers receive prompt care and that incident data are accurately recorded for subsequent investigation. Lessons from occupational health literature emphasize preparedness through scenario-based practice, not theoretical instruction alone. Medical readiness reflects organizational care for its workforce, reinforcing trust and morale. The ultimate measure of professionalism is not how accidents are avoided, but how swiftly and compassionately response occurs when they do.

## Investigating Incidents: From Data to Understanding

Incident investigation is the process through which organizations learn from failure. Its objective is not to assign blame but to reveal causation—mechanical, procedural, and human. The methodology integrates techniques from safety analysis frameworks such as root cause analysis, fault tree evaluation, and deviation analysis. Each begins with evidence collection: photographs, witness statements, and physical observations. Investigators reconstruct events chronologically to identify both direct and underlying factors. A near miss deserves the same analytical rigor as a serious injury, for it reveals vulnerabilities before consequences manifest. A mature safety organization treats every incident as a data point in a larger system. Investigations conclude with actionable recommendations—changes in design, procedure, or training that prevent recurrence. The strength of an investigation lies not in its documentation but in its implementation. When lessons are captured and shared, an incident becomes a teacher rather than a statistic. Over time, consistent analysis builds a culture of learning, transforming isolated failures into collective wisdom.

## Root Cause Analysis and the Chain of Causation

Root cause analysis explores the layers beneath an incident's surface. While an event may appear to result from human error, deeper investigation often uncovers systemic weaknesses—unclear procedures, inadequate supervision, or design flaws. The concept of the chain of causation describes how multiple factors align to produce failure. Removing a single link can prevent repetition. Root cause analysis follows structured methods: defining the problem, gathering data, identifying contributing factors, and distinguishing between immediate and fundamental causes. For instance, a fall from scaffolding may trace back to a missing guardrail, but the root cause may involve a deficient inspection process or unclear accountability. Corrective actions that address only the symptom achieve temporary relief; those that confront systemic issues yield sustainable safety. The analytical process demands objectivity, patience, and open dialogue among all involved. By uncovering root causes rather than scapegoats, organizations honor the principle that every accident is preventable if understood deeply enough.

## Emergency Response Coordination and Recovery

Response coordination transforms planning into action. When an emergency unfolds, leadership must integrate personnel, equipment, and external resources into a unified effort. The incident command system provides the structure to manage this complexity. Within it, roles are defined—incident commander, safety officer, communications lead, and logistics coordinator—ensuring no critical task is overlooked. Coordination also depends on clear situational awareness: responders must understand evolving hazards and adjust tactics accordingly. Once the immediate threat is contained, attention shifts to recovery. Re-establishing safe operations, accounting for personnel, and stabilizing the site require meticulous organization. Recovery is more than repair; it is restoration of confidence and continuity. Effective recovery also includes post-incident debriefs, allowing responders to reflect and refine procedures. Coordination succeeds when the transition from emergency to normal operations is smooth, documented, and instructive. Every response, well executed, reinforces readiness for the next.

## Psychological Resilience and Crisis Management

Emergencies test not only systems but human endurance. Psychological resilience becomes a decisive factor in maintaining order under stress. Crisis management recognizes that fear, confusion, and fatigue can impair decision-making as severely as physical hazards. Leaders must project calm authority, providing clear instructions and reassurance. Training that incorporates stress exposure—simulated alarms, timed drills, and decision-making under pressure—builds cognitive resilience. Support systems such as peer counseling or employee assistance programs aid recovery after traumatic events. Organizations that ignore the mental dimension of crisis response risk long-term consequences such as anxiety, absenteeism, or diminished trust. Resilient teams emerge when individuals feel valued, informed, and supported. In construction environments where emergencies can escalate rapidly, maintaining psychological stability is as vital as maintaining equipment. Preparedness, communication, and empathy combine to form the emotional armor that enables effective crisis management.

## Continuous Improvement in Emergency Management

Emergency management evolves through reflection and refinement. After every drill or real event, organizations conduct after-action reviews to evaluate performance and identify gaps. Lessons learned must feed directly into revised procedures, training updates, and design modifications. The philosophy mirrors that of quality management: plan, do, check, and act. Data from incident reports, inspection findings, and response evaluations guide decisions for resource allocation and policy enhancement. Continuous improvement also involves benchmarking against industry best practices and regulatory changes. True maturity in emergency management is achieved when learning becomes habitual and improvement continuous. Each revision of an emergency plan marks progress from reactive compliance toward proactive excellence. The organization that learns systematically from its experiences transforms uncertainty into capability, ensuring that readiness is not an aspiration but a proven strength.

## Download Audio File and Flashcards

# Chapter 3: Safety Program Development, Implementation, and Sustainment

## The Strategic Purpose of a Safety Program

A safety program in construction is not a binder of rules; it is an evolving framework that defines how an organization protects people, property, and reputation. At its core lies the recognition that safety is a managerial function equal to cost, schedule, and quality. Strategic intent begins with leadership endorsement—public, persistent, and personal. When executives declare that safety performance defines business performance, the message cascades throughout the company. A structured program transforms ideals into measurable systems: policy statements evolve into procedures, and procedures generate behavior. A successful program balances prevention, compliance, and culture. It anticipates hazards through planning, establishes accountability through documentation, and reinforces expectations through verification. Its structure mirrors a management system—policy, planning, implementation, measurement, and review—ensuring that improvement is cyclical rather than reactive. The strategic purpose is not simply regulatory adherence; it is the creation of an environment where every decision considers safety implications. A mature safety program thus becomes the organization's conscience, guiding decisions from boardroom strategy to field execution, embedding responsibility into every role.

## Management Leadership and Organizational Commitment

Effective safety programs begin with leadership that demonstrates visible commitment. Managers influence behavior not through mandates alone but by modeling consistency. When leadership attends safety meetings, reviews incident data, and allocates resources without hesitation, the workforce perceives that safety is integral to success. OSHA 3885 emphasizes management leadership as the first principle of a robust system: define roles, establish accountability, and remove barriers to participation. Leadership commitment also manifests through budget allocation—funding for training, protective equipment, and competent supervision. Policies must articulate the expectation that production never supersedes safety. A signed corporate statement should reinforce this, serving both as a promise and a performance contract. Supervisors translate this vision into daily routines: pre-task planning, inspection, and open communication. True commitment is tested not during audits but during production pressure. Organizations that halt unsafe work despite cost impact prove credibility. Over time, leadership consistency builds trust, transforming safety from a directive into a shared belief. Sustained commitment is measured not only by reduced incidents but by improved engagement and morale, confirming that leadership example remains the strongest preventive control.

## Worker Participation and Safety Ownership

Participation converts compliance into culture. Workers who contribute to hazard identification, procedure review, and program evaluation become owners of the system rather than subjects of enforcement. Modern safety philosophy rejects the outdated notion that safety is imposed from above; instead, it is co-created. Daily toolbox meetings, suggestion systems, and joint safety committees are practical expressions of involvement. Worker input captures site-specific insight often invisible to management—temporary hazards, behavioral trends, or process inefficiencies. OSHA's recommended practices highlight that active participation enhances problem-solving and innovation. Empowerment also strengthens reporting: when workers trust that feedback leads to improvement rather than blame, near-miss data becomes abundant and accurate. Inclusion must extend to all workforce levels and subcontract tiers, ensuring a uniform standard across diverse employers. Training in communication and observation supports participation by giving workers the vocabulary to describe risks precisely. The psychological shift from "being watched" to "being responsible" marks the evolution of a mature program. Participation transforms safety from an administrative expectation into a daily dialogue between policy and practice.

## Program Planning and Documentation

Planning defines the skeleton of a safety program. Without written structure, consistency cannot exist across shifts or projects. Program planning begins with identifying applicable regulations, client requirements, and organizational objectives. Documents such as safety manuals, site-specific plans, and task-hazard analyses form the governance framework. The Construction Safety Manual demonstrates that each contractor must prepare a comprehensive Site-Specific Safety & Health Program reviewed before mobilization. Documentation clarifies accountability—who inspects, who reports, who approves corrective actions. Yet documentation must remain functional, not bureaucratic. The best programs use concise templates that guide rather than overwhelm. Effective plans define performance metrics: inspection frequency, observation counts, training hours, and leading indicators. They also establish revision cycles to integrate lessons learned. Documentation is evidence of intent and a foundation for improvement. In audits, clear records demonstrate diligence; in emergencies, they guide decisions. The discipline of writing, reviewing, and updating embeds foresight into daily operations. Planning transforms uncertainty into direction, ensuring that safety objectives remain visible, measurable, and actionable throughout the project lifecycle.

## Implementation Through Training and Competence

No program survives poor implementation. Execution begins with ensuring that everyone understands and can perform their safety responsibilities. Competence is built through orientation, task-specific instruction, and periodic reinforcement. Training must blend knowledge with practice—workers must demonstrate proficiency, not merely attendance. Programs described in the Sample Construction Safety System emphasize structured orientation for all personnel, complemented by refresher sessions and supervisory coaching. Effective implementation uses layered communication: written procedures, visual signage, and verbal reinforcement. Supervisors act as translators, turning policy into practical behavior. Implementation also requires monitoring compliance through observation and feedback. When training is treated as an investment rather than expense, organizations reap productivity gains and reduced turnover. Competence is verified by performance in the field—safe habits sustained under pressure. Implementation transforms design into action, converting the abstract principles of the safety plan into lived behavior. A program succeeds only when every individual, from apprentice to project manager, demonstrates competence aligned with risk level and task complexity.

## Hazard Prevention, Control, and Program Integration

Prevention is the operational heart of safety management. It begins with systematic hazard identification, evaluation, and elimination. Control measures follow the hierarchy—engineering first, then administrative, and finally personal protective equipment. The Construction Safety Manual and OSHA 3885 both underscore proactive "find and fix" approaches. Integrating prevention across departments ensures that safety considerations inform procurement, scheduling, and design decisions. For example, procurement should evaluate equipment for ergonomic suitability and safety features; project scheduling should allow sufficient time for safe installation sequences. Integration requires collaboration among supervisors, engineers, and safety professionals so that controls are embedded in workflow rather than appended later. Continuous site inspections verify effectiveness, and corrective actions are documented through standard forms. Prevention also involves recognizing human factors—fatigue, complacency, and cognitive overload—and managing them through rotation and rest planning. A prevention-driven program views safety not as constraint but as enabler: by reducing interruptions, it supports efficiency. True integration ensures that hazard control becomes inseparable from operational excellence.

## Safety Performance Measurement and Auditing

Measurement translates effort into evidence. A safety program must track both leading and lagging indicators to evaluate effectiveness. Traditional metrics—recordable incidents, lost-time rates—reveal outcomes but not causes. Modern programs expand measurement to include behavioral observations, near-miss reports, corrective-action closure rates, and training participation. Auditing, as described in *The Art of Safety Auditing*, verifies that systems function as intended. Internal audits provide self-assessment; external audits offer impartial validation. Each audit cycle should conclude with a management review identifying trends and systemic weaknesses. Data integrity is critical—false reporting undermines credibility and improvement. Dashboards and scorecards can visualize performance, but interpretation requires context; a decline in incident rate may reflect reduced reporting rather than genuine improvement. Audits must be constructive, emphasizing learning rather than punishment. Findings become the blueprint for refinement. When performance measurement is transparent, employees see progress, leadership sees accountability, and clients see reliability. Auditing thus transforms safety from assumption to verified assurance, reinforcing that what is measured is managed.

## Continuous Improvement and Corrective Action

Sustainment depends on the discipline of continuous improvement. After each audit, incident, or review, corrective actions must be defined, assigned, and verified. The principle mirrors quality management's "Plan–Do–Check–Act" cycle. Corrective actions eliminate root causes rather than symptoms, ensuring that lessons generate real change. Programs should maintain a tracking system that records status, responsibility, and completion dates. Management review meetings assess progress, prioritize resources, and adjust policy accordingly. Continuous improvement also involves benchmarking against industry peers and standards. The Construction Safety Program model advocates annual evaluations to verify alignment with current regulations and best practices. Improvement culture thrives where openness is encouraged—reporting problems is valued as contribution, not failure. The ability to learn quickly defines resilience. When improvement becomes habitual, safety performance evolves from compliance to innovation. The mature organization recognizes that perfection is unattainable but progress is mandatory, using each review as an opportunity to strengthen structure, culture, and capability.

## Cultural Integration and Behavioral Reinforcement

Culture sustains what systems begin. A program achieves permanence only when safety becomes instinctive behavior supported by shared values. Cultural integration requires alignment between organizational beliefs and daily conduct. Leaders must reinforce positive behaviors through recognition, not merely penalties. Behavioral safety observations, peer feedback, and visible rewards encourage repetition of safe habits. Culture also thrives on storytelling—sharing lessons from incidents and successes humanizes data, connecting policy to emotion. In construction environments, where transient workforces challenge consistency, cultural rituals such as start-of-shift talks and joint commitment pledges build identity. Communication must remain open and bilingual—speaking both technical and human language. The concept of "just culture" from auditing literature promotes fairness: accountability balanced with learning. When workers believe that reporting errors leads to improvement rather than punishment, engagement deepens. Cultural integration ensures that safety survives leadership transitions and project turnover. It transforms procedures into values, ensuring continuity through shared conviction rather than external enforcement.

## Sustaining the Program Through Leadership Renewal

Sustainment is the final test of maturity. Programs often fail not because of design flaws but due to fading attention. Sustaining success requires periodic leadership renewal—new champions who bring energy without discarding proven foundations. Mentorship pipelines should prepare supervisors and foremen to assume higher safety responsibilities. Formal succession planning ensures continuity of vision. Sustaining also involves technological adaptation: integrating digital inspection tools, mobile training apps, and real-time data analytics to maintain relevance. Annual strategic reviews align the safety program with emerging risks such as automation, mental health, and environmental sustainability. External recognition, such as participation in voluntary protection initiatives, reinforces accountability. Ultimately, sustainment depends on belief: that safety excellence is never finished. Each project becomes both application and laboratory, testing the program's resilience under new constraints. Through leadership renewal and adaptive learning, the safety program transcends documentation, evolving into a living system that endures beyond individual projects and continues to protect those who build the world.

# Chapter 4: Leadership, Communication, and Training

## The Essence of Safety Leadership

Safety leadership represents the capacity to influence people toward safer choices through conviction, trust, and example. It transcends management authority, relying instead on personal credibility and consistency of action. A leader's behavior communicates priorities more powerfully than any written policy. When supervisors stop work to correct unsafe acts, they demonstrate that protection of life outweighs production targets. The IOGP Safety Leadership Model emphasizes four attributes—vision, engagement, integrity, and care. Vision provides direction; engagement fosters participation; integrity builds confidence; and care humanizes authority. Leadership begins by recognizing that safety excellence is not achieved through compliance alone but through shared purpose. Construction environments, with their temporary teams and shifting conditions, demand leaders who can build unity quickly. Leadership must therefore be adaptive, empathetic, and persistent. It is measured not by titles but by influence—the number of people who choose to follow safe practices without coercion. True safety leadership converts responsibility into inspiration, ensuring that every decision reflects respect for human life.

## Transformational Leadership and Safety Culture

Transformational leadership aligns closely with the creation of positive safety culture. It focuses on motivating individuals through vision and shared values rather than control. Transformational leaders articulate clear goals, recognize achievements, and challenge teams to exceed minimum expectations. Studies summarized in *Improving Safety Culture* reveal that organizations led by transformational figures experience higher trust, lower incident rates, and greater openness in reporting. Such leaders communicate optimism and curiosity, framing safety as opportunity instead of constraint. They inspire commitment by linking safety performance to personal pride and collective success. In construction, this approach is particularly effective where diverse subcontractors must unite under one standard. The leader's consistency bridges cultural and linguistic gaps, converting policy into shared identity. Transformational leadership also nurtures psychological safety—the confidence to speak up about hazards without fear of reprisal. This emotional climate accelerates learning and accountability. When leadership stimulates intrinsic motivation rather than external pressure, safety becomes self-sustaining, embedded in the team's identity rather than dependent on supervision alone.

## Building Trust and Credibility

Building trust and credibility is essential in professional relationships, leadership roles, and service-oriented environments. Trust forms the foundation of effective communication, collaboration, and long-term success. Credibility refers to the perception that an individual or organization is reliable, competent, and honest. Without trust, even technically skilled professionals may struggle to influence or lead others. Establishing credibility requires consistent behavior, ethical conduct, and demonstrated expertise. Trust is not created instantly but developed over time through repeated positive interactions. A strong reputation enhances both personal and organizational effectiveness.

Honesty is a central element in building trust. Transparent communication about intentions, expectations, and limitations fosters confidence. When individuals admit mistakes and take responsibility, credibility increases rather than decreases. Concealing information or providing misleading statements damages relationships. Truthful dialogue encourages mutual respect and understanding. Consistency between words and actions reinforces reliability. Over time, integrity becomes a defining characteristic of professional identity. Honesty strengthens long-term partnerships.

Competence is equally important in establishing credibility. Demonstrating knowledge, skills, and sound judgment reassures others of reliability. Continuous learning ensures that professionals remain up to date with evolving standards. Delivering high-quality work consistently reinforces confidence. When expertise is evident, others are more willing to rely on guidance and decisions. Competence must be combined with humility to avoid arrogance. Seeking feedback and improvement reflects dedication to excellence. Skilled performance strengthens professional standing.

Consistency in behavior contributes significantly to trustworthiness. Predictable actions allow others to feel secure in interactions. Sudden changes in attitude or commitment may create uncertainty. Meeting deadlines and honoring commitments demonstrates responsibility. Reliability in small tasks builds confidence for larger responsibilities. Consistency also applies to ethical standards and professional boundaries. Stable conduct creates a sense of safety and dependability. Over time, consistency transforms into reputation.

Effective communication plays a vital role in building trust. Clear, respectful, and empathetic communication prevents misunderstandings. Active listening shows that concerns and perspectives are valued. Providing timely updates keeps stakeholders informed and engaged. Tone and body language influence how messages are received. Open dialogue encourages collaboration and shared decision-making. Addressing concerns promptly prevents escalation of conflict. Communication that is both honest and compassionate strengthens credibility.

Respect is a fundamental component of trust-building. Valuing diverse perspectives promotes inclusion and fairness. Demonstrating courtesy and professionalism fosters positive relationships. Acknowledging contributions reinforces a sense of belonging. Respect for confidentiality protects sensitive information. Cultural sensitivity ensures equitable treatment across diverse groups. When individuals feel respected, they are more likely to cooperate and engage. Mutual respect enhances teamwork and morale.

Accountability strengthens both trust and credibility. Accepting responsibility for decisions and outcomes demonstrates maturity. Transparent documentation and reporting support accountability. Addressing errors promptly prevents long-term damage. Avoiding blame-shifting preserves professional integrity. Accountability builds confidence in leadership and management. Reliable performance under scrutiny enhances reputation. Trust deepens when individuals consistently uphold their responsibilities.

Building trust also requires emotional intelligence. Understanding and managing emotions improves interpersonal interactions. Empathy allows professionals to connect with others' experiences and concerns. Self-awareness prevents reactive or defensive behavior. Emotional regulation supports calm and constructive dialogue. Recognizing nonverbal cues enhances communication effectiveness. Emotional intelligence fosters supportive and collaborative environments. Trust grows when interactions are thoughtful and balanced.

Leadership significantly influences trust within organizations. Leaders who model integrity inspire similar behavior in others. Transparent decision-making reduces suspicion and uncertainty. Fair distribution of responsibilities demonstrates justice and respect. Encouraging participation empowers team members.

Recognition of achievements builds motivation and loyalty. Leaders who remain approachable promote open communication. Strong leadership establishes a culture of trust.

Credibility also depends on ethical conduct. Adhering to professional standards reinforces integrity. Avoiding conflicts of interest protects objectivity. Confidentiality safeguards sensitive information. Ethical behavior must remain consistent even under pressure. Short-term gains achieved through unethical actions undermine long-term credibility. Upholding moral principles demonstrates reliability and professionalism. Ethical consistency strengthens public confidence.

Trust can be fragile and easily damaged. Broken promises or dishonesty may erode years of credibility. Rebuilding trust requires transparency, apology, and corrective action. Consistent efforts are necessary to restore confidence. Acknowledging the impact of mistakes shows respect for affected parties. Patience and accountability are essential during recovery. Prevention through proactive communication reduces risk of breaches. Maintaining trust demands ongoing vigilance.

Collaboration enhances trust when roles and expectations are clearly defined. Shared goals encourage cooperation and mutual support. Open discussion of challenges promotes collective problem-solving. Fair recognition of contributions strengthens morale. Conflict resolution conducted respectfully preserves relationships. Collaboration built on transparency fosters long-term success. Trust within teams enhances productivity and innovation. Collective credibility reflects shared commitment.

In conclusion, building trust and credibility requires honesty, competence, consistency, and ethical conduct. Clear communication and emotional intelligence strengthen professional relationships. Respect, accountability, and fairness reinforce reliability. Leadership plays a central role in shaping a culture of trust. While trust takes time to develop, it can be quickly undermined by misconduct. Continuous self-improvement and transparency maintain credibility. Ultimately, trust and credibility form the foundation for effective leadership, collaboration, and sustained professional success.

Trust is the invisible currency of leadership. Without it, even the most elaborate programs collapse into formality. Credibility arises from alignment between words and actions; every promise kept reinforces belief, every inconsistency erodes it. Workers evaluate credibility through observation: whether supervisors wear required protective gear, respond promptly to concerns, and maintain fairness in discipline. Trust cannot be demanded—it must be earned daily through transparency, competence, and respect. The *PGSL Leadership Framework* highlights that leaders gain influence by listening first. When workers feel heard, they reciprocate with honesty and cooperation. Admitting uncertainty or mistakes paradoxically strengthens trust, demonstrating authenticity. In crisis situations, trust determines whether instructions are followed instantly or questioned hesitantly. Leaders who cultivate credibility create emotional safety, allowing teams to focus on problem-solving rather than self-protection. In the transient culture of construction, where relationships form and dissolve rapidly, trust anchors stability. It transforms compliance from obligation into voluntary commitment, ensuring cooperation even in high-pressure circumstances.

## Effective Communication in Safety Management

Communication is the mechanism through which safety leadership operates. It is both the message and the medium—verbal, visual, and behavioral. Effective communication requires clarity, consistency, and empathy. Ambiguous instructions or technical jargon often lead to misunderstanding, especially among multilingual crews. Leaders must adapt their language to the audience, using plain words and visual reinforcement. *Improving Safety Culture* emphasizes the "listen-verify-act" sequence: leaders listen actively, verify comprehension, and act visibly on feedback. Daily briefings, toolbox talks, and shift handovers are moments of influence where tone and presence shape perception. Communication also includes nonverbal cues; posture, attention, and reaction convey respect or dismissal. Digital tools enhance communication but cannot replace interpersonal connection. Successful programs encourage two-way dialogue, not one-way instruction. Workers are more likely to comply with procedures they helped refine through discussion. Ultimately, communication in safety is not transmission of information but creation of understanding—transforming directives into shared meaning that guides collective action.

## Coaching, Mentoring, and Behavioral Influence

Coaching, mentoring, and behavioral influence are powerful strategies for professional growth, leadership development, and performance improvement. Although these concepts are closely related, each serves a distinct purpose in supporting individuals and teams. Coaching is typically goal-oriented and focused on improving specific skills or performance outcomes. Mentoring involves a broader developmental relationship that supports long-term personal and professional growth. Behavioral influence refers to the intentional shaping of attitudes, actions, and decision-making patterns through guidance and example. Together, these approaches foster learning, confidence, and sustained improvement. Understanding their differences and connections enhances effective leadership and organizational success.

Coaching is a structured process designed to help individuals achieve defined objectives. It often focuses on measurable performance outcomes such as skill enhancement, productivity, or leadership capability. Coaches use questioning techniques to stimulate reflection and problem-solving. Rather than giving direct answers, they guide individuals to discover solutions independently. This approach builds self-awareness and accountability. Feedback is a central component of effective coaching. Through continuous dialogue and evaluation, coaching promotes focused and practical development.

Mentoring differs from coaching in scope and duration. Mentoring relationships are typically long-term and developmental in nature. A mentor shares experience, wisdom, and professional insights with a less experienced individual. Unlike coaching, mentoring may not always focus on specific short-term goals. Instead, it emphasizes career guidance, personal growth, and professional identity. Mentors provide encouragement, advice, and perspective based on lived experience. Trust and mutual respect are foundational to successful mentoring relationships. Over time, mentoring fosters confidence and strategic thinking.

Behavioral influence involves shaping attitudes and actions through intentional strategies. Leaders and professionals influence behavior through communication, modeling, reinforcement, and feedback. Positive reinforcement encourages repetition of desired behaviors. Constructive correction guides improvement without discouragement. Behavioral influence relies heavily on credibility and trust. Individuals are more likely to adopt behaviors demonstrated by respected role models. Ethical influence prioritizes empowerment rather than manipulation. Responsible influence strengthens organizational culture and teamwork.

Effective coaching requires strong communication skills. Active listening allows the coach to understand challenges and motivations. Open-ended questions encourage deeper thinking and self-exploration. Clear and constructive feedback supports improvement. Empathy builds rapport and psychological safety. When individuals feel supported rather than judged, they are more open to change. Coaching conversations should remain focused on goals and progress. Structured dialogue enhances measurable outcomes.

Mentoring relationships thrive on mutual commitment and trust. Regular interaction allows for meaningful guidance and reflection. Mentors share not only technical expertise but also professional values. They help mentees navigate challenges, opportunities, and career decisions. Encouraging independence is a key mentoring responsibility. Mentors must avoid creating dependency. Instead, they aim to empower mentees to develop their own strengths and judgment. Long-term guidance fosters resilience and professional maturity.

Behavioral influence is closely connected to organizational culture. Leaders shape norms through their daily actions and decisions. Consistent modeling of ethical behavior sets clear expectations. Recognition of positive contributions reinforces desirable conduct. Clear policies and accountability structures maintain standards. Social influence within teams encourages alignment with shared goals. When influence is aligned with core values, it strengthens unity and motivation. Cultural consistency enhances performance and morale.

Feedback is central to coaching, mentoring, and behavioral influence. Constructive feedback identifies areas for improvement while acknowledging strengths. Timely feedback prevents minor issues from becoming persistent problems. Balanced feedback supports both performance and confidence. Individuals must also be encouraged to provide upward feedback. Two-way communication enhances mutual growth. Feedback should be specific, actionable, and respectful. Effective feedback loops sustain continuous development.

Emotional intelligence enhances effectiveness in all three approaches. Self-awareness allows coaches and mentors to manage personal biases. Empathy strengthens understanding of others' perspectives. Emotional regulation supports calm and constructive interactions. Social awareness improves influence and communication strategies. High emotional intelligence fosters trust and openness. Individuals respond positively to supportive and respectful guidance. Emotional competence strengthens relational impact.

Goal setting is an important aspect of coaching and behavioral change. Clear goals provide direction and measurable benchmarks. Collaborative goal development increases commitment. Progress tracking reinforces motivation and accountability. Adjusting goals as circumstances change maintains relevance. Achievable milestones build confidence. Structured planning enhances clarity and focus. Goal alignment with organizational objectives supports overall success.

Ethical considerations are essential in coaching and mentoring relationships. Confidentiality must be respected to maintain trust. Boundaries prevent conflicts of interest or favoritism. Guidance should empower rather than control. Manipulative influence undermines credibility and morale. Transparency strengthens integrity. Ethical conduct ensures that influence is constructive and respectful. Professional standards guide responsible practice.

Challenges may arise in coaching and mentoring processes. Resistance to feedback can slow progress. Mismatched expectations may create frustration. Time constraints may limit meaningful interaction. Addressing these challenges requires clear communication and adaptability. Establishing shared expectations at the outset reduces misunderstanding. Flexibility ensures relevance to evolving needs.

Continuous reflection improves effectiveness. Overcoming obstacles strengthens developmental relationships.

Coaching, mentoring, and behavioral influence contribute significantly to leadership development. Leaders who coach effectively empower their teams. Mentoring prepares future leaders through knowledge transfer and example. Positive influence shapes collaborative and ethical cultures. Organizations that prioritize these strategies experience higher engagement and retention. Developmental support fosters loyalty and performance excellence. Investment in people strengthens long-term sustainability. Growth-oriented environments encourage innovation and adaptability.

In conclusion, coaching, mentoring, and behavioral influence are interconnected strategies that promote growth and performance. Coaching focuses on targeted skill development and measurable outcomes. Mentoring supports broader, long-term professional development. Behavioral influence shapes attitudes and actions through modeling and reinforcement. Effective communication, emotional intelligence, and ethical conduct enhance all three approaches. Continuous feedback and goal alignment sustain progress. Together, these strategies build capable individuals, strong leaders, and resilient organizations.

Coaching and mentoring are leadership techniques that shape long-term behavior. A coach guides individuals to identify their strengths and limitations, offering constructive feedback that empowers growth. Mentoring extends this relationship, providing continuity and trust over time. Both methods move beyond supervision to collaboration. In the context of safety, coaching occurs in the field—correcting behaviors through dialogue rather than reprimand. The *IOGP Report 453* highlights that the most influential mentors use questioning rather than instruction, prompting reflection: "What could make this task safer?" This approach develops self-awareness and decision-making capacity. Mentorship also accelerates competency transfer, ensuring that experience is not lost through turnover. When senior workers invest in newcomers, knowledge becomes generational. Behavioral influence is subtle but profound; consistent reinforcement of safe acts creates habit strength, while recognition of improvement sustains motivation. Coaching and mentoring transform leadership from authority into partnership, cultivating independence within the workforce. Over time, the organization benefits from a self-correcting culture where safety consciousness becomes automatic.

## The Dynamics of Team Communication

Teams in construction settings function under intense time and environmental pressure, making clear communication essential for coordination and safety. A team's effectiveness depends on shared mental models—common understanding of goals, hazards, and procedures. Miscommunication is a frequent precursor to incidents; therefore, structured dialogue techniques such as call-back verification and closed-loop communication are critical. Leaders must ensure that messages are not only sent but received and understood. The *PGSL Book on Safety Leadership* notes that team resilience grows from inclusive discussion where all members feel authorized to voice concerns. Morning briefings and debriefs establish rhythm, while visual tools like hazard boards maintain shared awareness throughout the day. Teams also rely on informal communication networks—peer reminders, quick exchanges, and signals—that fill the gaps between formal systems. Encouraging openness within these networks enhances adaptability. Effective team communication blends discipline with spontaneity, ensuring information moves freely without distortion. When communication flows smoothly, coordination improves, risk perception sharpens, and collective vigilance becomes instinctive.

## Conflict Resolution and Psychological Safety

Conflict in safety management often arises not from defiance but from misunderstanding, stress, or competing priorities. Constructive resolution requires empathy and emotional intelligence. Leaders must distinguish between behavior that challenges authority and behavior that challenges unsafe norms. Open dialogue transforms confrontation into learning. The *Improving Safety Culture* text stresses that psychological safety—the belief that one can speak up without humiliation or retaliation—is a prerequisite for honesty. When workers fear reprisal, silence conceals hazards. Conflict resolution begins with listening, validating emotion, and redirecting focus to shared goals. Neutral facilitation and consistent policy enforcement prevent escalation. Cultural awareness is also essential; communication styles vary across backgrounds, and sensitivity prevents misinterpretation. A psychologically safe environment fosters innovation because employees feel free to suggest improvements. In such climates, disagreements become opportunities for refinement rather than division. Leaders who master conflict resolution convert tension into progress, ensuring that diverse perspectives strengthen safety rather than fragment it.

## Training as a Strategic Investment

Training represents the operational link between leadership vision and workforce capability. It is not a compliance exercise but a strategic investment in competence and confidence. Effective safety training follows the principles of adult learning: relevance, interaction, and reinforcement. Adults learn best when material connects directly to their experience and when participation replaces passive reception. *TheLib.net safety training guide* emphasizes blended methods—classroom explanation, field demonstration, and scenario-based simulation—to translate theory into practice. Each training module must include measurable outcomes: what the learner can perform safely afterward. Evaluation through observation, quizzes, and practical demonstration confirms retention. Refresher sessions sustain memory and adapt content to evolving hazards. Training must also extend to supervisors, who require advanced instruction in leadership, investigation, and communication. Investment in training yields measurable returns: fewer incidents, lower turnover, and higher morale. A strategic training program ensures that competence is continuous, not event-based, reinforcing the idea that knowledge is the first control measure against risk.

## The Feedback Loop and Continuous Learning

The feedback loop and continuous learning are essential concepts in both technical systems and professional development. A feedback loop is a process in which the output of a system is returned as input to influence future performance. This mechanism allows systems to self-correct, stabilize, and improve over time. Continuous learning refers to the ongoing process of acquiring knowledge, reflecting on experiences, and adapting practices accordingly. Together, these concepts promote growth, efficiency, and resilience. Whether applied in engineering, healthcare, education, or organizational management, feedback and learning drive improvement. Understanding their principles supports long-term success and sustainability.

A feedback loop consists of several key components: input, process, output, and feedback. The input represents the initial information or stimulus entering the system. The process transforms this input into an output through specific actions or operations. The output is the result produced by the system. Feedback occurs when information about the output is measured and returned to adjust the input or process. This cycle continues repeatedly to refine performance. Effective feedback loops rely on accurate measurement and timely response.

There are two primary types of feedback loops: positive and negative feedback. Negative feedback reduces deviations and promotes stability within a system. For example, when performance declines, corrective action restores balance. Positive feedback amplifies changes and can accelerate growth or transformation. While positive feedback may enhance innovation, it must be controlled to prevent instability. Both types serve important functions depending on the context. Balanced systems often incorporate both forms of feedback. Proper management ensures productive outcomes.

In organizational settings, feedback loops improve performance and decision-making. Regular performance evaluations provide insights into strengths and weaknesses. Constructive feedback encourages employees to refine skills and behaviors. Leaders who actively seek feedback demonstrate openness and accountability. Transparent communication strengthens trust within teams. When feedback is ignored, errors may repeat and performance may decline. Integrating feedback into planning enhances adaptability. Organizations that value feedback foster a culture of improvement.

In professional practice, continuous learning ensures competence and relevance. Fields such as healthcare, engineering, and technology evolve rapidly. Professionals must update knowledge to maintain high standards of care or service. Continuing education, workshops, and certifications support skill development. Reflective practice allows individuals to analyze their experiences critically. Learning from both successes and failures strengthens expertise. Commitment to lifelong learning enhances credibility and performance. Continuous growth supports ethical and professional responsibility.

The feedback loop plays a crucial role in education and skill acquisition. Learners receive information about their performance through assessments and evaluations. Constructive feedback helps identify areas needing improvement. Timely responses enable learners to correct mistakes before they become habits. Encouragement boosts motivation and confidence. Repetition combined with feedback refines competence. Educators who provide clear guidance enhance learning outcomes. The cycle of practice and feedback strengthens mastery.

Technology has enhanced feedback systems in modern environments. Digital monitoring tools provide real-time data on performance and outcomes. In industrial systems, sensors detect variations and trigger automatic adjustments. In education, online platforms offer immediate assessment results. Data analytics helps identify trends and improvement areas. Automated systems reduce delays in corrective action. However, human interpretation remains essential for meaningful learning. Combining technology with reflective analysis maximizes effectiveness.

Continuous learning is closely connected to adaptability. In rapidly changing environments, rigid systems struggle to survive. Learning organizations encourage experimentation and innovation. Employees are motivated to share ideas and insights. Mistakes are viewed as opportunities for growth rather than failure. Adaptable systems respond quickly to new challenges. Feedback informs strategic adjustments. This dynamic approach ensures resilience and competitiveness.

Effective feedback must be clear, specific, and constructive. Vague or overly critical feedback may discourage improvement. Balanced feedback highlights strengths while addressing weaknesses. Respectful communication fosters openness and collaboration. Listening to feedback is as important as giving it. Mutual exchange strengthens professional relationships. Feedback should focus on behaviors and outcomes rather than personal traits. A supportive environment enhances receptiveness and growth.

Reflection is a central component of continuous learning. Reflective practice involves examining experiences to gain insight and improve future performance. This process requires honesty and self-awareness. Structured reflection may involve journaling, peer discussion, or mentorship. Reflection transforms experience into knowledge. Without reflection, learning opportunities may be lost. Continuous improvement depends on thoughtful analysis. Reflection strengthens professional maturity.

Leadership plays a vital role in establishing effective feedback loops. Leaders set expectations for accountability and growth. By modeling openness to feedback, they encourage others to do the same. Supportive leadership promotes psychological safety within teams. Clear goals and measurable outcomes guide improvement efforts. Recognition of progress reinforces motivation. Leadership commitment ensures feedback processes remain active. Strong guidance fosters a culture of continuous development.

Challenges may arise in implementing feedback and learning systems. Resistance to change can hinder progress. Fear of criticism may prevent honest communication. Limited resources may restrict training opportunities. Overemphasis on negative feedback can damage morale. Addressing these challenges requires supportive policies and encouragement. Building trust enhances participation. Effective systems balance accountability with compassion.

Continuous learning contributes to long-term success and sustainability. It enhances problem-solving abilities and innovation. Organizations that prioritize learning remain competitive. Individuals who pursue ongoing development maintain professional relevance. Feedback-driven systems reduce errors and inefficiencies. Adaptation to changing conditions becomes more manageable. Growth becomes an ongoing process rather than a one-time effort. Commitment to learning ensures steady advancement.

In conclusion, the feedback loop and continuous learning are interconnected processes that promote improvement and resilience. Feedback provides the information necessary to adjust performance and refine outcomes. Continuous learning transforms experiences into meaningful growth. Together, they support adaptability, competence, and innovation. Effective communication, reflection, and leadership strengthen these processes. Overcoming challenges requires commitment and openness. Embracing feedback and lifelong learning leads to sustained excellence in both individuals and organizations.

Sustained excellence in leadership and communication depends on feedback—the mechanism that transforms experience into improvement. Feedback must be timely, specific, and balanced, acknowledging success while addressing gaps. In safety management, feedback flows both upward and downward; workers provide insight on practicality, while leaders share performance results and lessons learned. Formal systems such as observation cards or digital reporting tools capture trends, but informal conversations often produce the most candid feedback. The *IOGP Safety Leadership Report* emphasizes learning reviews as alternatives to blame-based investigations, encouraging curiosity instead of punishment. Continuous learning occurs when feedback triggers adaptation—procedures refined, training updated, leadership styles adjusted. Organizations that celebrate feedback as participation rather than criticism develop agility. They respond to weak signals before crises arise. The feedback loop thus becomes a cycle of resilience: observation, reflection, correction, and innovation. Through this dynamic exchange, leadership remains relevant and safety culture remains alive.

## Sustaining Leadership Through Culture and Example

Leadership longevity depends on culture—shared values that endure beyond individual personalities. Sustaining leadership means institutionalizing its behaviors through systems, rituals, and stories. Every new leader should inherit a model of conduct rather than reinvent it. The *Improving Safety Culture Guide* highlights that rituals such as safety moments, recognition ceremonies, and storytelling preserve lessons and reinforce priorities. Sustained leadership also requires humility: the willingness to keep learning and adapting. As technology, workforce demographics, and societal expectations evolve, so must leadership methods. Mentorship programs ensure continuity, while transparent communication maintains credibility through transition. Leaders who cultivate successors rather than followers leave a lasting legacy. Ultimately, leadership is sustained through example—consistent demonstration that safety is nonnegotiable, compassion is strength, and integrity defines success. When such values become habitual across generations, leadership ceases to depend on individuals; it becomes embedded in the organization's identity, ensuring enduring protection for everyone involved in the work.

Sustaining leadership through culture and example is essential for long-term organizational success and stability. Leadership is not only about authority or decision-making power, but also about influence, values, and consistent behavior. A strong organizational culture reinforces shared beliefs, expectations, and standards of conduct. When leaders align their actions with these values, they strengthen credibility and trust. Culture acts as a guiding framework that shapes daily interactions and long-term strategy. Leadership sustained by culture becomes resilient during times of change or crisis. By modeling desired behaviors, leaders create environments where excellence can thrive.

Organizational culture represents the collective norms, attitudes, and behaviors that define how work is accomplished. It influences communication styles, decision-making processes, and employee engagement. A healthy culture promotes accountability, collaboration, and respect. Leaders play a central role in shaping and reinforcing cultural values. When culture aligns with mission and vision, it creates unity and purpose. Employees are more motivated when they understand shared goals. Strong culture provides consistency even when leadership transitions occur.

Leading by example is one of the most powerful tools for sustaining leadership. Employees closely observe how leaders act under pressure and during routine tasks. Consistency between words and actions builds credibility. If leaders promote integrity but act dishonestly, trust quickly erodes. Demonstrating commitment, professionalism, and ethical behavior sets a clear standard. Modeling positive attitudes influences team morale. Behavior often speaks louder than formal policies. Authentic leadership inspires long-term loyalty and engagement.

Trust is a fundamental component of culture-driven leadership. Leaders who communicate openly and honestly foster psychological safety. Employees feel confident sharing ideas and concerns without fear of retaliation. Transparent decision-making reduces suspicion and misunderstanding. Accountability strengthens mutual respect. Trust allows teams to function effectively even in uncertain conditions. When leaders demonstrate fairness and empathy, trust deepens. A culture of trust enhances productivity and collaboration.

Consistency in leadership behavior reinforces cultural stability. Sudden or unpredictable changes in standards can create confusion and insecurity. Leaders who uphold clear expectations promote clarity and alignment. Regular communication of values strengthens understanding. Recognition of behaviors that reflect organizational principles reinforces consistency. Stable leadership reduces uncertainty during organizational challenges. Predictability in conduct builds confidence in direction. Consistency transforms leadership from authority into influence.

Empowerment is another critical element in sustaining leadership through culture. Leaders who encourage participation create a sense of ownership among team members. Delegating responsibility fosters confidence and skill development. When employees feel valued, they contribute more effectively. Empowerment strengthens innovation and adaptability. Leaders who trust their teams build reciprocal trust. Shared leadership encourages accountability at all levels. Empowered cultures remain strong even when leaders are absent.

Ethical conduct serves as the foundation for lasting leadership. Leaders who uphold fairness, transparency, and integrity set a moral example. Ethical behavior must remain consistent, especially during difficult decisions. Favoritism or dishonesty damages credibility and morale. Clear ethical standards guide behavior across the organization. Ethical leadership reduces conflicts and builds respect. A strong moral framework supports sustainable success. Integrity ensures that leadership influence remains positive.

Communication plays a vital role in shaping and sustaining culture. Leaders must articulate vision, expectations, and goals clearly. Regular updates keep teams aligned and informed. Active listening demonstrates respect for diverse perspectives. Addressing concerns promptly prevents misinformation. Open communication fosters inclusion and engagement. Feedback loops allow continuous improvement. Effective dialogue strengthens shared understanding.

Adaptability is necessary for sustaining leadership in evolving environments. Organizational culture must remain flexible while preserving core values. Leaders who model resilience inspire confidence during change. Encouraging learning and innovation supports long-term relevance. Adaptable cultures respond constructively to challenges. Change managed through example reduces resistance. Leaders who remain composed during uncertainty reinforce stability. Growth-oriented cultures sustain momentum over time.

Recognition and reinforcement strengthen cultural alignment. Celebrating achievements that reflect organizational values motivates continued excellence. Positive reinforcement shapes desired behaviors. Recognition programs demonstrate appreciation and respect. Leaders who acknowledge contributions foster loyalty. Fair and transparent recognition builds trust. Reinforcement aligns daily actions with strategic objectives. Appreciation enhances morale and engagement.

Mentorship and role modeling contribute to cultural continuity. Experienced leaders pass knowledge and values to emerging leaders. Mentorship ensures that organizational principles remain intact across generations. Observing ethical and competent leadership shapes professional identity. Succession planning strengthens long-term stability. Cultural consistency depends on leadership development. Role models inspire commitment and professionalism. Investing in leadership growth preserves institutional strength.

Challenges to sustaining leadership may arise from rapid growth, crisis, or external pressures. In such situations, strong culture provides guidance and stability. Leaders must remain visible and consistent during adversity. Clear communication prevents confusion and fear. Reinforcing shared values strengthens unity. Addressing issues promptly maintains credibility. Resilient cultures overcome challenges more effectively. Leadership grounded in culture withstands disruption.

Continuous reflection and evaluation support sustainable leadership. Leaders should assess whether actions align with stated values. Seeking feedback enhances accountability and growth. Adjustments may be necessary to maintain cultural relevance. Commitment to improvement strengthens authenticity. Reflective practice prevents complacency. Sustainable leadership evolves while preserving core principles. Ongoing learning reinforces adaptability and strength.

In conclusion, sustaining leadership through culture and example requires alignment between values and actions. Organizational culture provides structure and direction for consistent behavior. Leaders who model integrity, accountability, and empathy strengthen trust and unity. Empowerment and communication reinforce engagement and resilience. Ethical conduct ensures long-term credibility. Adaptability and mentorship preserve continuity during change. By leading through example and nurturing a strong culture, leaders create lasting influence and sustainable success.

# Chapter 5 – Practice Questions

## Questions (1-100)

**Q1. Which control method is the most effective in the hierarchy of controls for eliminating fall hazards at a construction site?**

A. Personal fall arrest systems
B. Guardrails and toe boards
C. Engineering design changes to remove the hazard
D. Administrative rotation of workers

**Q2. When conducting a Job Hazard Analysis (JHA), what is the first step in the process?**

A. Identify potential hazards
B. Select the job to be analyzed
C. Develop control measures
D. Train employees on findings

**Q3. Which of the following best defines a confined space in construction work?**

A. A space that is fully enclosed with continuous airflow
B. Any space that contains mechanical equipment
C. A tunnel or trench less than 4 feet deep
D. A space large enough to enter, with limited entry or exit, not designed for continuous occupancy

**Q4. Which control measure should be implemented first when workers are exposed to excessive noise levels on-site?**

A. Require earplugs
B. Substitute quieter equipment
C. Rotate workers to different areas
D. Enforce shorter shifts

**Q5. A worker is assigned to grind metal in a fabrication area. Sparks and noise levels are both high. Which control is most appropriate to address both hazards?**

A. Isolation booth with sound insulation
B. PPE including ear defenders and fire-retardant clothing
C. Engineering ventilation and spark shields
D. Job rotation every hour

**Q6. Which hazard control should be prioritized when dealing with exposure to silica dust during concrete cutting?**

A. Providing respirators to workers
B. Implementing wet cutting methods
C. Installing warning signage
D. Scheduling work during off-hours

**Q7. During excavation, which condition most commonly leads to trench collapse?**

A. Poor soil classification and lack of shoring
B. Inadequate lighting
C. Worker fatigue
D. Overhead utility interference

**Q8. Which of the following is the most reliable indicator of an electrical hazard on-site?**

A. The presence of nearby metal scaffolding
B. Use of ungrounded temporary wiring
C. Cables running overhead

D. High humidity levels

**Q9. What is the purpose of conducting air monitoring in confined spaces before entry?**

A. To measure worker performance under pressure
B. To ensure adequate oxygen and detect toxic gases
C. To assess noise and vibration levels
D. To locate ventilation ducts

**Q10. When using cranes near power lines, what is the minimum action required before beginning work?**

A. Assign a signal person only
B. Ground all metal parts of the crane
C. Determine the minimum approach distance and de-energize if necessary
D. Limit boom extension to half the distance

**Q11. In the hierarchy of controls, which is the least effective method for hazard management?**

A. Elimination
B. Substitution
C. Engineering control
D. Personal protective equipment

**Q12. What is the most critical reason to maintain good housekeeping on a construction site?**

A. To reduce insurance costs
B. To improve site aesthetics
C. To prevent slips, trips, and material fires
D. To demonstrate management control

**Q13. Which control is most appropriate to reduce risk from vibration exposure during power tool use?**

A. Gloves with padded palms
B. Vibration-damped tools and task rotation
C. Warm-up exercises for hands
D. Periodic tool calibration

**Q14. What type of ventilation is most effective for controlling welding fumes at the source?**

A. Natural airflow through open doors
B. Local exhaust ventilation (LEV)
C. General dilution ventilation
D. Periodic air changes every hour

**Q15. A concrete worker handling form-release oil experiences skin irritation. What immediate control should be implemented?**

A. Switch to water-based release agents
B. Require face shields
C. Increase work-rest cycles
D. Use longer-handled brushes

**Q16. Which hazard is best controlled using a permit-to-work system?**

A. Routine painting activities
B. Confined space entry and hot work
C. Housekeeping and waste disposal
D. Equipment calibration tasks

**Q17. When performing hot work in a construction zone, what is the best pre-control action?**

A. Ensure fire extinguishers are nearby
B. Remove combustible materials and issue a hot work permit
C. Assign a fire watch only
D. Check the weather forecast

**Q18. Which factor contributes most to struck-by incidents involving heavy equipment?**

A. Poor visibility and lack of spotters
B. Engine vibration
C. Hydraulic fluid leaks
D. Poor fuel efficiency

**Q19. Which of the following is a key element of a risk assessment record?**

A. Worker's age and experience
B. List of identified hazards and control measures
C. Names of site visitors
D. Copies of training certificates

**Q20. Which type of PPE is most critical for workers performing abrasive blasting?**

A. Safety glasses
B. Hard hats
C. Earplugs only
D. Full-face respirator with supplied air

**Q21. What is the primary purpose of a safety data sheet (SDS)?**

A. To document employee training
B. To provide hazard information and handling instructions for chemicals
C. To track inventory levels
D. To record incident investigations

**Q22. During manual handling, what factor most contributes to musculoskeletal injury risk?**

A. Wearing gloves
B. Load weight, posture, and frequency
C. Ambient noise levels
D. Type of footwear

**Q23. A worker is exposed to loud intermittent noise from impact tools. What control method reduces this specific hazard most effectively?**

A. Sound barriers or enclosures around equipment
B. Ear muffs for all workers nearby
C. Limiting tool usage to daytime
D. Installing reflective sound panels

**Q24. What should be verified before entering a confined space with a history of chemical storage?**

A. Lighting sufficiency
B. Atmospheric testing for oxygen, flammables, and toxins
C. Worker hydration level
D. Proximity to exit ladders

**Q25. When using hand and power tools, what preventive measure reduces kickback injuries?**

A. Wearing double gloves
B. Using properly maintained and sharp tools
C. Applying extra downward force
D. Disabling tool guards for comfort

**Q26. What is the safest approach when dealing with overhead loads lifted by cranes on a construction site?**

A. Walk under suspended loads quickly
B. Allow workers to guide loads by hand
C. Use tag lines to control loads and establish exclusion zones
D. Move equipment near the lifting path

**Q27. What should be done first when a worker reports dizziness after entering a confined space?**

A. Administer oxygen immediately

B. Have the worker exit the space and test the atmosphere
C. Provide water and rest inside the space
D. Continue work but monitor the situation

**Q28. Which hazard is most associated with unprotected floor openings on upper levels of buildings?**
A. Falling objects striking workers below
B. Electrical arc flash
C. Equipment overheating
D. Exposure to chemical fumes

**Q29. When using scaffolding, which inspection frequency provides the highest level of safety compliance?**
A. Monthly visual inspection
B. Before first use, after alterations, and every 7 days thereafter
C. Only when defects are suspected
D. After project completion

**Q30. Which of the following best controls exposure to lead during renovation work?**
A. Post warning signs only
B. Dry sweeping dust daily
C. Use wet methods and HEPA-filtered vacuum systems
D. Rotate workers frequently

**Q31. What should be done before beginning excavation work near a suspected underground utility line?**
A. Proceed carefully using a backhoe
B. Mark and hand-dig within the tolerance zone after verifying utility locations
C. Assume all lines are inactive
D. Ignore markings if plans are outdated

**Q32. A welder experiences minor electric shock while working. What immediate step should the safety officer take?**
A. Replace welding leads with new ones immediately
B. Inspect equipment for grounding and insulation defects
C. Increase output voltage to improve efficiency
D. Resume work after a brief break

**Q33. Which practice reduces the risk of hand injuries when using power tools?**
A. Wearing loose gloves for flexibility
B. Keeping guards in place and using two-handed controls when available
C. Working quickly to minimize exposure
D. Disabling safety interlocks

**Q34. When working with chemical adhesives, what storage condition ensures safety?**
A. Store in sealed containers away from heat sources
B. Keep containers open for ventilation
C. Mix with other chemicals to neutralize fumes
D. Store under direct sunlight for drying

**Q35. Which engineering control minimizes fall risks on open-sided platforms?**
A. Visual caution tape
B. Standard guardrails with mid-rails and toe boards
C. Safety posters and signs
D. Mandatory use of harnesses only

**Q36. Which type of control best protects against repetitive strain injuries from tool vibration?**
A. Administrative training sessions
B. Anti-vibration tool design and ergonomic scheduling

C. Extra PPE layers on hands
D. Cooling breaks every two hours

**Q37. During mechanical lifting operations, what condition most commonly results in equipment tip-over?**
A. Operator fatigue
B. Short duration lifts
C. Lack of seatbelts
D. Overloading or uneven ground conditions

**Q38. What control should be implemented to prevent fires when storing flammable liquids on-site?**
A. Storing near heaters for warmth
B. Keeping containers closed and grounded in ventilated areas
C. Mixing liquids to save space
D. Storing inside occupied trailers

**Q39. What is the safest lifting method for heavy loads handled manually?**
A. Twist while lifting for better balance
B. Lift with legs, keeping back straight and load close to body
C. Use arms only to minimize leg strain
D. Hold load far from torso to see better

**Q40. Which action helps prevent material storage collapse?**
A. Leveling storage areas and maintaining load stability
B. Stacking materials to maximum height
C. Removing labels to reduce clutter
D. Leaving materials unsecured for quick access

**Q41. What hazard arises when welding in poorly ventilated areas?**
A. UV light reflection
B. Toxic fume accumulation
C. Equipment overheating
D. Reduced weld penetration

**Q42. When is respiratory protection required on construction sites?**
A. When airborne contaminants exceed permissible exposure limits
B. Whenever dust is visible
C. When comfort masks are available
D. Only during chemical mixing tasks

**Q43. What is the main control objective of the Globally Harmonized System (GHS)?**
A. Regulate international trade in chemicals
B. Standardize chemical classification and labeling for safety communication
C. Replace OSHA regulations
D. Track material quantities on-site

**Q44. Which factor most contributes to hand-arm vibration syndrome (HAVS)?**
A. Short tool bursts at low vibration
B. Prolonged exposure to high-frequency vibrating tools
C. Occasional manual handling
D. Use of insulated gloves

**Q45. What control reduces risks from working near moving vehicles on-site?**
A. Allow free movement to increase efficiency
B. Use flashing lights only
C. Install physical barriers and assign spotters for reversing
D. Require drivers to sound horns constantly

**Q46. Which hazard is associated with improper use of powder-actuated tools?**
A. Overheating only
B. Projectile penetration and ricochet injuries
C. Electrical short circuits
D. Reduced productivity
**Q47. What safety control should be verified before crane operations begin each shift?**
A. Tire pressure of nearby vehicles
B. Crane's load charts and mechanical inspection completion
C. Weather forecast accuracy
D. Number of workers on-site
**Q48. During demolition, which condition most increases risk of building collapse?**
A. Use of hand tools
B. Removing structural supports prematurely
C. Presence of dust clouds
D. High worker turnover
**Q49. Which step ensures safety when handling asbestos-containing materials (ACM)?**
A. Break materials to fit disposal bags
B. Wet materials and seal them in labeled containers
C. Store dry and uncovered for quick removal
D. Transport with regular waste
**Q50. Which hazard is primarily associated with prolonged noise exposure above 85 dBA?**
A. Skin irritation
B. Muscular fatigue
C. Eye strain
D. Noise-induced hearing loss
**Q51. What is the primary control for protecting workers from falling objects during overhead work?**
A. Personal fall arrest systems
B. Toe boards, debris nets, and barricaded drop zones
C. Hard hats only
D. Rotating workers below frequently
**Q52. What is the most important step before performing hot work in an area with potential flammable vapor presence?**
A. Open all nearby doors and windows
B. Test the atmosphere with a calibrated gas detector
C. Assign two fire watchers for the duration
D. Keep water hoses available nearby
**Q53. When using a ladder on uneven ground, what method provides the most stable base?**
A. Lean the ladder against a wall more steeply
B. Have another worker hold the ladder continuously
C. Use a ladder leveler or secure base pads
D. Place blocks under one leg of the ladder
**Q54. Which type of hazard is created when compressed gas cylinders are stored horizontally?**
A. Fire hazard only
B. Explosion risk from weakened cylinder valves
C. Trip hazard only
D. Excessive moisture accumulation
**Q55. What is the key purpose of conducting daily pre-use equipment inspections?**
A. To meet documentation requirements only

B. To identify defects before they result in failure or injury
C. To increase productivity by tracking tool use
D. To verify workers know how to use equipment

**Q56. When handling hazardous materials, why is proper labeling critical?**
A. It reduces storage space requirements
B. It ensures clear identification and hazard communication
C. It simplifies waste disposal only
D. It replaces training requirements

**Q57. Which type of footwear best prevents puncture injuries from nails on construction sites?**
A. Steel-toe boots with puncture-resistant midsoles
B. Rubber boots
C. Lightweight sneakers
D. Non-slip dress shoes

**Q58. Which hazard most commonly results from inadequate housekeeping on multi-level structures?**
A. Electric shock
B. Fire and trip hazards from accumulated debris
C. Radiation exposure
D. Inadequate illumination

**Q59. What key factor must be verified before allowing work at height using scaffolding?**
A. The weather forecast for rain
B. Competent person inspection and load capacity compliance
C. Crew's lunch schedule
D. Scaffold material color

**Q60. What hazard exists when high-pressure hydraulic hoses are improperly maintained?**
A. Minor leaks with no major risk
B. Increased air humidity
C. Injection injuries and potential fluid burns
D. Reduced equipment mobility

**Q61. Which factor should be prioritized when selecting respiratory protection for chemical exposure?**
A. Weight of the respirator
B. Type and concentration of airborne contaminant
C. Worker's age and gender
D. Equipment color

**Q62. What action should be taken if a safety monitoring alarm for gas detection is triggered?**
A. Continue work until confirmation from management
B. Evacuate immediately and investigate the cause
C. Turn off the alarm and resume work
D. Move only non-essential personnel out

**Q63. What is the main hazard of using electrical extension cords beyond their rated capacity?**
A. Reduced lighting quality
B. Overheating and potential fire
C. Static electricity buildup
D. Premature insulation discoloration

**Q64. What control reduces injury risk when handling rebar vertically on construction sites?**
A. Capping exposed ends and proper securing
B. Painting rebar ends for visibility
C. Increasing work pace to finish early
D. Lifting rebar manually without equipment

**Q65. Why should electrical panels be kept clear of stored materials?**
A. To allow quick access during emergencies
B. To reduce cleaning requirements
C. To improve room aesthetics
D. To provide extra storage space nearby
**Q66. Which hazard occurs when oxygen cylinders are stored near fuel gases?**
A. Noise pollution
B. Explosion due to accelerated combustion
C. Low oxygen concentration
D. Static shock
**Q67. What is the main advantage of implementing safety through design in construction planning?**
A. Reduces need for supervision
B. Eliminates hazards before construction begins
C. Speeds up procurement
D. Minimizes tool purchases
**Q68. What is the safest method for controlling dust during demolition work?**
A. Periodic sweeping
B. Conducting demolition at night
C. Using industrial fans to blow dust away
D. Continuous water spraying or misting systems
**Q69. When performing overhead drilling, which PPE is most critical to prevent injuries?**
A. Standard safety glasses
B. Full-face shield and hard hat combination
C. Ear plugs only
D. Welding goggles
**Q70. Which principle should guide prioritizing hazards during risk assessment?**
A. Severity and likelihood of occurrence
B. Worker preference
C. Supervisor's opinion
D. Order of detection
**Q71. When selecting PPE, what factor ensures it provides effective protection?**
A. Matching color with uniform
B. Proper fit, maintenance, and compatibility with other equipment
C. Using the heaviest model available
D. Purchasing in bulk for savings
**Q72. What engineering control can prevent slips on wet concrete surfaces?**
A. Posting caution signs
B. Installing surface texturing or anti-slip coatings
C. Restricting work hours
D. Providing larger footwear
**Q73. Which hazard control minimizes crane load swing during lifting operations?**
A. Tag lines and steady lift speed
B. Reduced boom length
C. High-speed hoisting
D. Manual push by workers
**Q74. What should a supervisor do immediately after a near-miss involving hazardous energy release?**
A. Resume work if no one was hurt
B. Investigate the cause and implement corrective actions

C. Dismiss the report as minor
D. Focus on paperwork completion
**Q75. Which control reduces exposure to airborne biological contaminants on construction sites?**
A. Avoiding water use entirely
B. Increasing rest breaks
C. Wearing reflective vests
D. Maintaining proper drainage and ventilation
**Q76. What immediate control should be used when workers must perform grinding operations indoors?**
A. Provide respirators only
B. Use local exhaust ventilation and spark guards
C. Rotate workers every hour
D. Keep windows open for air circulation
**Q77. What is the most effective control when handling materials that emit harmful vapors?**
A. Wearing gloves only
B. Allowing vapors to dissipate naturally
C. Using closed transfer systems and proper ventilation
D. Scheduling work at night
**Q78. Which factor should be verified before using a portable generator on-site?**
A. Its paint color
B. Proper grounding and GFCI protection
C. Operator's age
D. Distance to lunch area
**Q79. Which step reduces ergonomic strain during repetitive bricklaying tasks?**
A. Use adjustable scaffolding and material platforms
B. Work faster to finish early
C. Avoid rest breaks
D. Lift bricks one at a time from ground level
**Q80. What control best prevents heat stress among workers wearing impermeable protective suits?**
A. Shorter meal breaks
B. Cooling vests and scheduled hydration intervals
C. Dark-colored uniforms
D. Working during midday hours
**Q81. Which hazard is most associated with working around unguarded rotating equipment?**
A. Noise irritation
B. Entanglement injuries and amputations
C. Vibration discomfort only
D. Temporary hearing loss
**Q82. Which measure best controls chemical splash hazards during mixing operations?**
A. Wearing long-sleeve cotton shirts
B. Mixing faster to reduce exposure time
C. Standing further from the workstation
D. Using face shields, gloves, and chemical-resistant aprons
**Q83. Which control helps reduce risk of crane boom collapse during lifting operations?**
A. Conducting daily pre-use inspections and adhering to load charts
B. Operating beyond rated capacity for efficiency
C. Lifting loads without rigging checks
D. Using visual judgment only to estimate load weight
**Q84. What procedure ensures safety before entering a manhole for maintenance?**

A. Assume it is safe if recently used
B. Test atmosphere and ventilate continuously
C. Lower tools before air testing
D. Skip monitoring if work is brief
**Q85. Which factor contributes most to slip hazards on stairways at construction sites?**
A. Open design with wide treads
B. Poor lighting and accumulation of debris
C. Handrail presence
D. High step uniformity
**Q86. What immediate control prevents trench cave-ins during excavation?**
A. Posting warning signs
B. Limiting worker entry
C. Shoring, sloping, or benching based on soil classification
D. Increasing work speed
**Q87. What is the purpose of grounding portable electric tools?**
A. Improve motor speed
B. Prevent electric shock from fault currents
C. Increase tool durability
D. Reduce noise emission
**Q88. Which control best minimizes risk when handling sharp-edged sheet metal?**
A. Wearing loose cotton gloves
B. Stacking sheets haphazardly
C. Avoiding gloves for better grip
D. Using cut-resistant gloves and edge covers
**Q89. What hazard results from using high-pressure air for cleaning surfaces?**
A. Noise only
B. Injection injuries and airborne debris hazards
C. Static shock
D. Minor discomfort
**Q90. Which hazard is most common when working around suspended scaffolds?**
A. Falling due to lack of guardrails or fall arrest systems
B. Heat exposure
C. Limited visibility
D. Poor communication
**Q91. What control should be implemented when using fuel-powered equipment indoors?**
A. Keep windows closed to reduce noise
B. Provide adequate ventilation or use exhaust extraction systems
C. Place fans nearby to circulate air
D. Limit fuel tank size only
**Q92. What control prevents backflow contamination in temporary water lines?**
A. Open drain valves daily
B. Install backflow prevention devices
C. Disconnect hoses after each use
D. Use smaller hoses only
**Q93. Which hazard is most significant when handling asbestos insulation?**
A. Fire risk
B. Contact burns
C. Inhalation of airborne fibers

D. Water contamination

**Q94. When lifting with a mobile crane on uneven terrain, what step ensures stability?**

A. Shortening the boom fully

B. Using properly leveled outriggers and mats

C. Operating at maximum boom speed

D. Ignoring minor tilts

**Q95. What hazard is introduced by improper lighting in confined spaces?**

A. Static buildup

B. Trips, missteps, and electrical shock from wet conditions

C. Noise reflection

D. Increased humidity

**Q96. What control minimizes risk during welding near flammable materials?**

A. Using plastic tarps to catch sparks

B. Increasing welding speed

C. Wearing lighter gloves

D. Clearing the area and using fire-resistant blankets

**Q97. What control prevents contact with live electrical conductors?**

A. Insulating exposed conductors and locking out circuits

B. Wearing metal jewelry for conductivity

C. Working with wet gloves

D. Increasing voltage gradually

**Q98. What is the best practice for storing oxygen and acetylene cylinders together?**

A. Store upright with at least 20 feet separation or noncombustible barrier

B. Store side-by-side for convenience

C. Lay them flat to prevent tipping

D. Keep them under a workbench

**Q99. What hazard arises from mixing incompatible chemicals?**

A. Mild odor release

B. Violent reaction producing toxic gases or heat

C. Reduced chemical potency

D. Increased evaporation rate

**Q100. What control ensures safe material hoisting using ropes or slings?**

A. Use damaged slings for lightweight loads

B. Inspect slings before each use and use proper hitch configuration

C. Increase lift speed to reduce swing

D. Attach multiple loads simultaneously

---

## Answer keys and Explanations (1-100)

**Q1. Answer: C**

**Explanation**: Eliminating the hazard through engineering design is the most effective control because it removes the risk entirely. This distinguishes it from guardrails and PPE, which only mitigate the hazard. Administrative measures are least effective since they rely on consistent human behavior rather than structural elimination.

**Q2. Answer: B**

**Explanation**: The JHA process begins by selecting the specific job to analyze so that each step can be systematically broken down. This distinguishes it from identifying hazards, which occurs after the job is chosen. Developing controls and training occur later, once the hazards and their causes have been fully understood.

**Q3. Answer: D**

**Explanation**: A confined space allows worker entry but has limited openings and is not intended for continuous occupancy. This differentiates it from standard tunnels or mechanical rooms, which may be designed for occupancy. Continuous airflow does not change the confined classification.

**Q4. Answer: B**

**Explanation**: Substitution with quieter equipment removes the noise source, which aligns with the higher levels of the control hierarchy. PPE and administrative actions are reactive and less reliable since they depend on consistent usage and compliance. Eliminating or substituting hazards offers long-term noise exposure reduction.

**Q5. Answer: C**

**Explanation**: Engineering controls, like spark shields and localized ventilation, directly reduce exposure to heat, sparks, and airborne metals. This is superior to PPE-only methods that depend on user behavior. Job rotation or isolation are supplementary but not primary solutions for dual physical hazards.

**Q6. Answer: B**

**Explanation**: Wet cutting minimizes dust generation at its source, effectively reducing airborne silica. This surpasses PPE, which only protects the individual, and signage or scheduling, which do not control exposure. Eliminating or suppressing the dust at generation is the most effective method.

**Q7. Answer: A**

**Explanation**: Trench collapses typically result from unstable soil conditions and failure to shore or slope the trench. Lighting and fatigue can influence safety but do not directly cause structural failure. Overhead utilities are different hazards unrelated to trench wall stability.

**Q8. Answer: B**

**Explanation**: Ungrounded wiring directly increases the likelihood of shock and electrocution. While overhead cables and humidity can elevate risk, grounding ensures circuit safety. Metal scaffolding can transmit current, but without ungrounded systems, the risk remains manageable.

**Q9. Answer: B**

**Explanation**: Air monitoring ensures safe oxygen concentration and identifies the presence of toxic or flammable gases. This distinguishes it from environmental comfort testing such as noise measurement. The focus is on life-threatening atmospheric conditions before worker entry.

**Q10. Answer: C**

**Explanation**: OSHA requires calculating the safe approach distance and contacting the utility to de-energize or shield lines. Merely grounding or using a signal person cannot prevent electrical arcs. Reducing boom extension does not ensure compliance or full protection.

**Q11. Answer: D**

**Explanation**: PPE relies on individual compliance and provides protection only after exposure occurs, making it the least reliable. Higher-order controls modify or remove the hazard entirely. Engineering and administrative actions address risks proactively, unlike PPE which is purely reactive.

**Q12. Answer: C**
**Explanation**: Effective housekeeping eliminates obstructions, reduces ignition sources, and enhances safe movement. While aesthetics and administrative benefits exist, the core value is direct injury and fire prevention. It's a proactive control addressing multiple hazards simultaneously.
**Q13. Answer: B**
**Explanation**: Vibration-damped tools limit transmitted energy and job rotation minimizes exposure duration. Gloves and calibration offer minor benefits but do not mitigate vibration intensity. Engineering and administrative combinations offer the most effective control balance for this hazard.
**Q14. Answer: B**
**Explanation**: LEV captures contaminants directly at the point of generation, preventing their spread. Natural or dilution ventilation only reduces overall concentration without addressing immediate worker exposure. Direct fume capture is crucial for welder respiratory protection.
**Q15. Answer: A**
**Explanation**: Substituting hazardous solvent-based oils with less toxic, water-based alternatives removes the source of irritation. PPE and procedural adjustments reduce contact but do not eliminate the cause. Product substitution effectively lowers health risk while maintaining operational efficiency.
**Q16. Answer: B**
**Explanation**: Permit-to-work systems regulate high-risk, non-routine tasks like hot work and confined space entry through authorization and control. Routine tasks do not warrant this formal system. The permit ensures hazards are identified and controlled before activity begins.
**Q17. Answer: B**
**Explanation**: Eliminating combustibles before hot work and formal authorization through a permit system minimizes ignition risk. Fire extinguishers and watchers are reactive, not preventive. Proactive removal and procedural approval ensure comprehensive control.
**Q18. Answer: A**
**Explanation**: Limited visibility and the absence of spotters lead to vehicle-human collisions. While mechanical issues affect performance, they rarely cause struck-by events. Spotter coordination and visual signaling directly prevent these incidents.
**Q19. Answer: B**
**Explanation**: A risk assessment record documents hazards, their associated risks, and selected control measures. Other details like age or training are administrative and not part of the assessment. The record focuses on identifying and controlling risk factors systematically.
**Q20. Answer: D**
**Explanation**: Abrasive blasting generates fine particulates and rebound materials, making respiratory protection essential. Safety glasses or hard hats provide limited coverage, and earplugs do not prevent inhalation hazards. Supplied-air respirators ensure comprehensive respiratory and facial protection.
**Q21. Answer: B**
**Explanation**: The SDS informs workers about chemical hazards, exposure limits, and safe handling practices. It is distinct from inventory or training records. Accessible SDSs enable proper PPE selection and emergency response, supporting hazard communication compliance.
**Q22. Answer: B**
**Explanation**: The combination of excessive load, awkward posture, and repetition directly increases strain and fatigue. Gloves or footwear aid grip and comfort but do not address biomechanical stress. Proper ergonomics and lifting methods reduce risk effectively.
**Q23. Answer: A**
**Explanation**: Physical barriers or acoustic enclosures directly absorb and isolate the noise source. PPE remains necessary but secondary. Administrative adjustments like scheduling offer minor reduction without addressing the actual emission.

**Q24. Answer: B**

**Explanation**: Atmospheric testing ensures safe entry conditions by checking for oxygen sufficiency and absence of hazardous vapors. Lighting and access features support safety but do not identify invisible dangers. This test is a mandatory prerequisite before confined entry.

**Q25. Answer: B**

**Explanation**: Well-maintained, sharp tools cut efficiently and reduce the likelihood of snagging or kickback. Forcing or disabling guards increases hazard potential. Maintenance and correct usage prevent tool-related hand injuries effectively.

**Q26. Answer: C**

**Explanation**: Using tag lines provides controlled load movement and prevents swing, while exclusion zones ensure no personnel are under the suspended load. This combination eliminates the most severe struck-by and crush risks. Allowing manual guidance or crossing under suspended loads directly violates safety protocols designed to protect life and property.

**Q27. Answer: B**

**Explanation**: Immediate evacuation followed by atmospheric testing ensures the worker is removed from a potentially toxic or oxygen-deficient environment. Administering oxygen or continuing work without analysis may worsen exposure. Proper testing identifies the contaminant source, allowing control actions before re-entry.

**Q28. Answer: A**

**Explanation**: Unprotected openings create fall risks for both workers above and below, primarily from falling objects. Electrical or chemical hazards are separate issues. Properly covering or guarding openings prevents both fall-through and debris hazards, making it a critical control on elevated work sites.

**Q29. Answer: B**

**Explanation**: Regular inspections ensure structural integrity is maintained throughout construction. Performing checks before use, after any modification, and weekly prevents failures caused by environmental or operational wear. Less frequent or reactive inspections may overlook critical hazards, leading to falls or collapses.

**Q30. Answer: C**

**Explanation**: Wet methods and HEPA-filtered vacuums prevent dust from becoming airborne, providing the most effective engineering control. Posting warnings and worker rotation only reduce exposure opportunities but do not eliminate the source. Dry sweeping increases inhalation risk by re-suspending toxic particles.

**Q31. Answer: B**

**Explanation**: Locating utilities precisely and hand-digging within the tolerance area minimizes damage risk. Mechanical digging without confirmation often leads to strikes causing electrocution or gas leaks. Assuming inactivity or ignoring maps is unsafe and violates standard excavation safety requirements.

**Q32. Answer: B**

**Explanation**: A shock event indicates potential grounding or insulation failure. Conducting an inspection isolates and corrects faults before work resumes. Replacing components or increasing voltage without analysis can worsen the hazard, while continuing work endangers the operator.

**Q33. Answer: B**

**Explanation**: Two-hand controls and intact guards prevent accidental activation and contact with moving parts. Loose gloves or tampering with interlocks increase entanglement risks. Tool control features are designed to stop unintentional triggering, ensuring maximum safety during operation.

**Q34. Answer: A**

**Explanation**: Properly sealed, temperature-controlled storage prevents vapor buildup and ignition. Open or mixed chemicals can create toxic or explosive reactions. Heat and sunlight accelerate chemical breakdown, raising both fire and exposure risks, so cool storage is essential.

**Q35. Answer: B**
**Explanation**: Guardrails with mid-rails and toe boards physically block falls and objects from dropping below. Signs or tape provide warnings but lack physical restraint. Harnesses are supplemental controls but do not prevent initial slip or edge contact, making guardrails primary.
**Q36. Answer: B**
**Explanation**: Anti-vibration designs directly reduce transmitted force, while ergonomic scheduling limits duration of exposure. Training or cooling breaks alone don't mitigate physical vibration intensity. Combining engineering and scheduling ensures sustainable protection from chronic musculoskeletal disorders.
**Q37. Answer: D**
**Explanation**: Overloading and unstable terrain shift the center of gravity, causing tipping. Operator fatigue contributes to human error but not direct mechanical imbalance. Stable ground assessment and adherence to rated capacities prevent this high-risk hazard effectively.
**Q38. Answer: B**
**Explanation**: Closed and grounded containers prevent vapor release and static ignition. Adequate ventilation disperses flammable vapors, reducing explosion potential. Storing near heat or in enclosed trailers raises ignition and health risks due to confined vapor accumulation.
**Q39. Answer: B**
**Explanation**: Bending knees and keeping the load close maintains spinal alignment, preventing strain. Twisting or lifting with arms alone overexerts small muscle groups. Maintaining a stable posture distributes load evenly, minimizing back and ligament injuries.
**Q40. Answer: A**
**Explanation**: Level surfaces and stable stacking prevent shifting and collapse, especially under vibration or wind. Overstacking and unsecured materials compromise integrity. Organized storage reduces potential energy release, avoiding struck-by or crush injuries.
**Q41. Answer: B**
**Explanation**: Inadequate ventilation traps metal oxides and gases, leading to respiratory illness or poisoning. Equipment performance is rarely affected, but worker exposure rises significantly. Ventilation ensures safe oxygen levels and removes harmful airborne contaminants.
**Q42. Answer: A**
**Explanation**: Respiratory protection is mandated once contaminant concentrations exceed regulatory limits. Visual dust alone does not determine requirement, nor do comfort masks suffice. Monitoring results dictate protective needs for all operations exceeding safe exposure thresholds.
**Q43. Answer: B**
**Explanation**: GHS ensures consistent hazard communication through standardized symbols, labels, and data sheets. It doesn't regulate trade or replace laws; rather, it enhances comprehension across industries. This improves worker understanding and handling of chemical hazards globally.
**Q44. Answer: B**
**Explanation**: Extended high-frequency tool use transmits harmful energy into nerves and vessels, leading to HAVS. Short, controlled bursts or manual handling do not reach harmful thresholds. Insulated gloves protect thermally but provide minimal vibration damping capability.
**Q45. Answer: C**
**Explanation**: Barriers create separation between workers and vehicles, and spotters enhance rear visibility. Horn use and flashing lights help awareness but cannot substitute physical or procedural separation. Coordinated control minimizes run-over and collision incidents.
**Q46. Answer: B**

**Explanation**: Powder-actuated tools operate explosively, propelling fasteners that can ricochet or penetrate unintended surfaces. Overheating and electrical issues are minimal concerns. Proper training and material verification are critical for preventing severe impact injuries.

**Q47. Answer: B**

**Explanation**: Confirming mechanical readiness and consulting load charts ensure lifting operations remain within safe limits. Weather affects conditions but does not replace required equipment checks. Consistent inspection prevents catastrophic equipment failure and ensures compliance.

**Q48. Answer: B**

**Explanation**: Prematurely removing supports destabilizes structures, causing unplanned collapses. Dust and labor changes pose secondary risks but not immediate failure triggers. Proper sequencing and engineering assessment prevent collapse-related fatalities during demolition activities.

**Q49. Answer: B**

**Explanation**: Wetting minimizes airborne fiber release, and sealed labeling prevents contamination during transport. Breaking or mixing asbestos releases hazardous fibers, violating handling regulations. Proper containment is essential for worker and environmental protection.

**Q50. Answer: D**

**Explanation**: Prolonged exposure to high decibel levels damages inner ear cells, causing irreversible hearing loss. Other listed conditions are unrelated to sound intensity. Implementing engineering and administrative controls along with hearing protection maintains auditory health effectively.

**Q51. Answer: B**

**Explanation**: Installing toe boards and debris nets prevents materials from falling, while barricading drop zones restricts entry into hazardous areas. Hard hats protect against smaller debris but not large falling objects. Engineering controls combined with exclusion zones ensure comprehensive protection from overhead hazards.

**Q52. Answer: B**

**Explanation**: Testing for flammable vapors with a gas detector ensures conditions are safe before ignition sources are introduced. Fire watchers and ventilation support safety but cannot substitute atmospheric testing. Identifying and confirming a non-explosive environment is essential to prevent fire or explosion incidents.

**Q53. Answer: C**

**Explanation**: Ladder levelers or base pads provide engineered stability that prevents slippage or tipping. Having someone hold the ladder offers temporary support but is unreliable. Using improvised blocks creates instability and increases accident risk, making levelers the safest option.

**Q54. Answer: B**

**Explanation**: Storing gas cylinders horizontally can damage or stress valves, causing potential explosion or uncontrolled release. Upright storage ensures valve protection and stability. Fire and trip hazards may exist but are secondary; pressure integrity is the primary safety concern.

**Q55. Answer: B**

**Explanation**: Daily inspections reveal early signs of wear, preventing accidents caused by mechanical failure. Documentation and productivity are secondary benefits. Detecting and correcting issues proactively ensures worker safety and operational reliability before hazards occur.

**Q56. Answer: B**

**Explanation**: Labeling provides instant hazard recognition and safety instructions, allowing correct handling and emergency response. It complements but does not replace worker training. Unlabeled materials cause confusion and increase risk of improper mixing or accidental exposure.

**Q57. Answer: A**

**Explanation**: Steel-toe boots with puncture-resistant midsoles protect both the foot top and sole against sharp objects. Rubber or athletic footwear lacks this reinforcement, allowing penetration from nails or rebar. Proper footwear is a vital PPE component against puncture hazards.

**Q58. Answer: B**

**Explanation**: Debris buildup leads to tripping and fire risks, especially near welding or hot work areas. Poor illumination contributes indirectly, but clutter directly obstructs paths and increases ignition potential. Regular cleanup eliminates immediate and secondary hazards efficiently.

**Q59. Answer: B**

**Explanation**: A competent person's inspection ensures structural stability and that load capacities are not exceeded. Weather checks are important but secondary. Verification confirms the platform can safely support workers and materials under expected loads and conditions.

**Q60. Answer: C**

**Explanation**: Damaged hydraulic hoses can inject fluid through skin at high velocity, causing severe internal injuries. Minor leaks quickly escalate under pressure. Regular inspection and maintenance prevent catastrophic hose failure and ensure worker protection during operation.

**Q61. Answer: B**

**Explanation**: The specific contaminant type and exposure level determine filter type and respirator class. Choosing based on weight or aesthetics ignores critical filtration requirements. Proper matching ensures effective protection from toxic gases or particulates.

**Q62. Answer: B**

**Explanation**: A triggered alarm indicates hazardous atmospheric conditions. Immediate evacuation ensures no one is exposed to potentially deadly gases. Ignoring or delaying response can result in asphyxiation or explosion. Investigation occurs only after the area is confirmed safe.

**Q63. Answer: B**

**Explanation**: Overloading cords causes excessive heat buildup, leading to insulation breakdown and possible fire. Static electricity and discoloration are minor effects. Ensuring proper current ratings prevents overheating and electrical hazards during prolonged tool use.

**Q64. Answer: A**

**Explanation**: Capping prevents impalement injuries, while secure storage stops accidental tipping. Visibility markings help identification but not protection. Engineering controls, such as rebar caps, directly prevent penetration injuries during material handling.

**Q65. Answer: A**

**Explanation**: Keeping access clear ensures emergency shutdowns and maintenance can be performed without delay. Obstructions hinder quick disconnection in critical incidents, increasing shock or fire risk. Clearance is both a regulatory and life-safety requirement.

**Q66. Answer: B**

**Explanation**: Oxygen accelerates combustion; if mixed with fuel gas leaks, explosion potential increases drastically. Separating gases and maintaining barriers eliminate this hazard. Other listed options do not reflect the severe reactive nature of oxygen-fuel proximity.

**Q67. Answer: B**

**Explanation**: Safety through design integrates hazard elimination during project planning, preventing risks before they reach the job site. Unlike later corrective actions, proactive design solutions reduce both injuries and costs by eliminating dangerous features early.

**Q68. Answer: D**

**Explanation**: Water suppression systems bind dust particles, preventing them from becoming airborne. Sweeping or fans disperse contaminants further, and night operations don't reduce exposure. Continuous misting maintains visibility and respiratory safety for all workers nearby.

**Q69. Answer: B**

**Explanation**: Overhead drilling exposes workers to falling debris and fine particles, requiring full facial and head protection. Safety glasses alone cover limited areas. The hard hat and shield together protect eyes, skin, and head from direct and rebound hazards.

**Q70. Answer: A**

**Explanation**: Risk is determined by evaluating potential severity and probability. Prioritizing based on these ensures resources focus on the most significant threats. Other methods rely on subjective judgment, which may overlook critical dangers requiring immediate control.

**Q71. Answer: B**

**Explanation**: PPE must fit correctly and function with other equipment like respirators or helmets. Poorly fitted or incompatible gear fails to protect adequately. Cost or color considerations cannot outweigh comfort and performance alignment for user safety.

**Q72. Answer: B**

**Explanation**: Textured finishes or anti-slip coatings increase surface friction, reducing slip potential significantly. Signage warns but doesn't prevent falls. Engineering solutions provide long-term safety improvements that remain effective under wet or oily conditions.

**Q73. Answer: A**

**Explanation**: Tag lines and controlled motion maintain stability and prevent uncontrolled swinging. Reducing boom length may limit range but not swing. Manual pushing is unsafe. Smooth lifts and guided lines ensure balance and load predictability during hoisting.

**Q74. Answer: B**

**Explanation**: Near-misses indicate process weaknesses that could lead to severe incidents. Investigating root causes enables targeted preventive measures. Ignoring such events forfeits learning opportunities and allows hazards to persist uncorrected in future operations.

**Q75. Answer: D**

**Explanation**: Drainage and ventilation reduce moisture accumulation and microbial growth, limiting airborne contamination. Reflective gear and rest schedules have no effect on bioaerosol levels. Environmental management is crucial for preventing respiratory infections and allergen buildup.

**Q76. Answer: B**

**Explanation**: Local exhaust ventilation captures fine dust and fumes directly at the source, preventing inhalation and buildup. Spark guards contain debris, reducing fire risk. Rotating workers or opening windows improves comfort but cannot effectively control contaminants generated during grinding operations.

**Q77. Answer: C**

**Explanation**: Closed systems prevent vapor release, and mechanical ventilation removes residual gases, protecting workers from inhalation hazards. Gloves alone provide no respiratory protection. Allowing natural dissipation or night work fails to control exposure effectively in confined or poorly ventilated areas.

**Q78. Answer: B**

**Explanation**: Grounding and GFCI protection prevent electric shock and equipment damage. Improperly grounded generators pose severe electrocution hazards. Color, operator age, or proximity to facilities are irrelevant to electrical safety and regulatory compliance.

**Q79. Answer: A**

**Explanation**: Adjustable scaffolding minimizes bending and reaching, maintaining neutral posture and reducing musculoskeletal stress. Increased pace or eliminating rest increases fatigue and injury risk. Proper height alignment supports long-term productivity and safety.

**Q80. Answer: B**

**Explanation**: Cooling vests regulate body temperature, and hydration maintains electrolyte balance. This combination prevents heat exhaustion and heat stroke effectively. Dark clothing absorbs more heat, and midday work increases exposure, making them unsafe practices for hot environments.

**Q81. Answer: B**
**Explanation**: Unguarded rotating parts can catch clothing, gloves, or hair, causing severe entanglement or amputation. Noise and vibration are secondary hazards. Machine guards are essential engineering controls that prevent direct contact with moving components.
**Q82. Answer: D**
**Explanation**: PPE designed for chemical resistance shields skin and eyes from corrosive splashes. Cotton absorbs chemicals and increases injury risk. Speed or distance adjustments do not provide consistent protection from unexpected spills or reactions.
**Q83. Answer: A**
**Explanation**: Inspections identify structural defects, while load charts verify capacity compliance. Exceeding limits or skipping rigging checks leads to catastrophic failures. Objective measurement and documentation ensure lifting operations remain within safe mechanical boundaries.
**Q84. Answer: B**
**Explanation**: Air testing identifies oxygen deficiency or toxic gas presence before entry, and continuous ventilation maintains safe conditions. Assuming safety or skipping testing disregards confined space standards. Proper procedures prevent asphyxiation and toxic exposure incidents.
**Q85. Answer: B**
**Explanation**: Dim lighting and debris accumulation obscure visibility and obstruct footing, leading to slips and trips. Handrails and consistent step design enhance safety, but without cleanliness and illumination, risks remain high. Regular maintenance prevents accidents effectively.
**Q86. Answer: C**
**Explanation**: Proper shoring or sloping stabilizes trench walls according to soil type. Signs and limited entry raise awareness but cannot prevent collapse. Structural reinforcement physically resists soil movement, ensuring worker protection during excavation operations.
**Q87. Answer: B**
**Explanation**: Grounding redirects fault current safely to earth, preventing shock injuries. It does not enhance performance or longevity. Ground connections are vital for all metallic tool casings used in damp or conductive environments to ensure user safety.
**Q88. Answer: D**
**Explanation**: Cut-resistant gloves prevent lacerations, and edge covers reduce accidental contact during movement. Cotton gloves and poor stacking create new hazards. Proper PPE and storage organization ensure safe handling of sharp materials.
**Q89. Answer: B**
**Explanation**: High-pressure air can penetrate skin or propel particles at dangerous velocities. This leads to injection injuries or eye trauma. Safe cleaning requires regulated air pressure, protective gear, and alternative cleaning methods like vacuum systems.
**Q90. Answer: A**
**Explanation**: The primary risk involves falls from height if guardrails or harnesses are missing. Environmental factors may complicate tasks, but lack of physical fall protection causes the most severe incidents. Consistent use of fall systems ensures worker safety.
**Q91. Answer: B**
**Explanation**: Exhaust gases contain carbon monoxide, requiring ventilation or extraction to prevent poisoning. Fans without ducting only disperse gases locally. Ventilation removes toxic fumes effectively, maintaining breathable air quality indoors.
**Q92. Answer: B**
**Explanation**: Backflow preventers stop contaminated water from reversing into clean systems. Simple draining or disconnection doesn't prevent reverse pressure events. Installation ensures potable water remains safe for all site operations.

**Q93. Answer: C**
**Explanation**: Disturbed asbestos releases microscopic fibers that cause lung disease upon inhalation. Fire and water hazards are secondary concerns. Containment, wetting, and PPE use are essential to prevent exposure during handling or removal.
**Q94. Answer: B**
**Explanation**: Outriggers distribute load evenly and mats stabilize footing. Uneven ground without leveling leads to tipping hazards. Shortening the boom changes reach but doesn't ensure base stability, making proper leveling the safest practice.
**Q95. Answer: B**
**Explanation**: Poor lighting obscures obstacles and standing water, leading to slips and electrical accidents. Noise and humidity are unrelated to visibility. Adequate illumination ensures safe navigation and accurate equipment handling in tight environments.
**Q96. Answer: D**
**Explanation**: Fire-resistant blankets and removing combustibles prevent ignition from sparks. Plastic tarps melt, creating secondary hazards. Proper preparation establishes a controlled safe zone and prevents potential fire spread during welding.
**Q97. Answer: A**
**Explanation**: Insulation and lockout/tagout procedures isolate live energy sources, preventing accidental contact. Metal objects and moisture increase conductivity and shock risk. Isolation and de-energization form the foundation of electrical safety practices.
**Q98. Answer: A**
**Explanation**: Physical separation or barriers prevent fire from spreading between oxygen and fuel gases. Horizontal or enclosed storage creates stability and ventilation problems. Compliance ensures explosive mixtures cannot develop under accident conditions.
**Q99. Answer: B**
**Explanation**: Incompatible chemicals react energetically, releasing heat or toxic vapors. Reduced potency or odor changes are minor compared to explosive reactions. Proper labeling and segregation prevent these dangerous incidents from occurring during handling.
**Q100. Answer: B**
**Explanation**: Pre-use inspection ensures sling integrity, and correct hitching balances the load to prevent drop or rotation. Damaged or overloaded slings cause failure. Safe lifting requires verified equipment and correct rigging geometry every time.

## Questions (101-200)

**Q101. Which control should be applied first when working near high-voltage power lines during construction?**

A. Installing warning tape on the ground
B. Posting a safety sign near the lines
C. Wearing insulated gloves only
D. Contacting the utility company to de-energize or relocate lines

**Q102. Which measure most effectively prevents chemical burns when handling strong acids or bases?**

A. Working in open air without protection
B. Rinsing hands frequently with water during use
C. Using ordinary gloves and aprons
D. Wearing chemical-resistant PPE and providing eyewash stations nearby

**Q103. What is the primary reason for implementing a lockout/tagout (LOTO) program on-site?**

A. To simplify electrical installation procedures
B. To reduce time during equipment maintenance
C. To replace the need for equipment guards
D. To prevent unexpected startup or energy release during servicing

**Q104. When working in extreme cold, which practice best prevents cold stress and frostbite?**

A. Using insulated moisture-wicking clothing and scheduled warm-up breaks
B. Working continuously without breaks
C. Wearing multiple cotton layers
D. Increasing caffeine intake

**Q105. What hazard is associated with poor illumination around stairways and corridors?**

A. Reduced air movement
B. Elevated noise levels
C. Shortened work duration
D. Increased trips, falls, and contact injuries

**Q106. What engineering control reduces exposure to diesel exhaust fumes in tunnels or enclosed sites?**

A. Running engines intermittently
B. Installing exhaust ventilation and catalytic converters
C. Using scented sprays to mask odors
D. Rotating workers every 30 minutes

**Q107. Which hazard control best protects workers during demolition near existing structures?**

A. Engineering assessments and controlled sequencing
B. Worker relocation without planning
C. Manual observation only
D. Increased demolition speed

**Q108. What precaution minimizes exposure to ultraviolet radiation during outdoor welding operations?**

A. Wearing cotton clothing and hats
B. Standing farther from the arc without PPE
C. Working under bright sunlight for visibility
D. Using UV-resistant shields, screens, and appropriate PPE

**Q109. What control is most important when operating forklifts on sloped surfaces?**

A. Driving slowly and keeping the load upgrade
B. Turning sideways on the slope
C. Increasing travel speed for momentum
D. Carrying extra counterweights

**Q110. What procedure ensures safety before servicing hydraulic systems?**
A. Cover components with cloth
B. Release stored energy and lock out pressure sources
C. Tighten all fittings
D. Depressurize after partial disassembly
**Q111. Which hazard control should be applied for work involving asbestos removal?**
A. Compressing waste to save space
B. Opening windows for air movement
C. Using wet removal methods and negative pressure enclosures
D. Dry sweeping debris to keep area clean
**Q112. What hazard arises when scaffolding planks extend excessively beyond their supports?**
A. Visual obstruction only
B. Increased risk of tipping or plank breakage
C. Reduced airflow
D. Noise hazard
**Q113. Which action minimizes risk when transporting flammable materials by vehicle?**
A. Transport with other incompatible chemicals
B. Leave containers open for ventilation
C. Use approved containers and secure upright during transport
D. Keep containers in passenger compartments
**Q114. What control prevents overexertion injuries when handling heavy loads manually?**
A. Skipping warm-up exercises
B. Working faster to reduce exposure time
C. Using mechanical lifting aids and team lifting for bulky items
D. Relying solely on personal strength
**Q115. What should be done before using compressed air for cleaning machinery parts?**
A. Clean without protective eyewear
B. Point nozzle close to the skin to test pressure
C. Use maximum pressure for efficiency
D. Reduce air pressure below 30 psi and use chip guards
**Q116. When working in areas with frequent heavy equipment movement, what safety measure is essential?**
A. Using mobile phones while signaling
B. Standing closer to machinery for better communication
C. Relying solely on sound alerts
D. Wearing high-visibility clothing and maintaining eye contact with operators
**Q117. What is the most critical factor when using personal fall arrest systems?**
A. Wearing only a body belt
B. Inspecting components before each use and ensuring proper anchorage
C. Connecting lanyard to movable structures
D. Using any available harness regardless of condition
**Q118. What hazard results from overloading aerial lifts?**
A. Equipment tipping or mechanical failure
B. Better reach performance
C. Reduced fuel consumption
D. Longer equipment lifespan
**Q119. Which control prevents static discharge during flammable liquid transfer?**
A. Ignoring grounding on short transfers

B. Pouring quickly to reduce exposure time
C. Using plastic containers only
D. Bonding and grounding containers before transfer

**Q120. What is the safest method for inspecting elevated steel structures?**
A. Conducting visual checks from ground level only
B. Leaning over edges without support
C. Climbing unprotected ladders for faster access
D. Using approved fall protection and secured anchorage points

**Q121. When is a spotter required during heavy equipment operation?**
A. Whenever operator visibility is limited or near personnel
B. Only when raining
C. When equipment is brand new
D. If the operator requests one

**Q122. Which action prevents accidental release of suspended loads?**
A. Increasing lift speed to minimize swing
B. Using defective slings for lighter weights
C. Inspecting rigging hardware before each use and securing hooks with safety latches
D. Allowing personnel under suspended loads

**Q123. Which factor increases risk of electrocution with power tools in wet conditions?**
A. Wearing rubber gloves
B. Keeping cords elevated
C. Using double-insulated tools with GFCI protection
D. Using ungrounded cords and damaged insulation

**Q124. Which hazard results from improper use of ladders near doorways?**
A. Ladder bending under weight
B. Workers being struck by opening doors
C. Ladder slippage due to vibration
D. Overreaching while on rungs

**Q125. What control prevents airborne exposure when cutting materials containing crystalline silica?**
A. Performing dry cutting for precision
B. Working downwind of others
C. Wearing only cloth masks
D. Using wet cutting and vacuum dust collection systems

**Q126. What precaution should be taken when using scaffolds in high wind conditions?**
A. Secure materials and delay work if wind exceeds safe limits
B. Anchor workers' belts to scaffold frames directly
C. Continue work with additional personnel for stability
D. Remove guardrails to reduce wind resistance

**Q127. Which hazard is most likely to occur when handling uncured concrete without protection?**
A. Thermal burns
B. Electrocution from rebar contact
C. Abrasions only
D. Chemical burns and skin irritation from caustic lime content

**Q128. What control measure is essential for safe hoisting operations near building edges?**
A. Relying solely on verbal communication
B. Employing barricades and using tag lines to stabilize loads
C. Using temporary fencing only
D. Allowing workers under suspended materials

**Q129. What step should be performed before using a new chemical product on-site?**
A. Store it near fuel sources for accessibility
B. Mix with other chemicals for testing
C. Review the Safety Data Sheet (SDS) and train workers
D. Open containers to smell test
**Q130. Which action helps prevent injury when moving long pipes through narrow passages?**
A. Using multiple workers and clear paths for maneuvering
B. Carrying pipes vertically for balance
C. Moving quickly to minimize time in confined spaces
D. Dragging pipes along the floor for control
**Q131. What hazard arises when compressed gas cylinders are exposed to direct sunlight for extended periods?**
A. Cylinder corrosion
B. Improved gas efficiency
C. Decreased gas pressure
D. Over-pressurization and potential explosion
**Q132. What immediate action prevents collapse of formwork during concrete placement?**
A. Allowing untrained laborers to adjust bracing
B. Checking form stability and bracing before and during placement
C. Speeding up pour rate
D. Waiting for concrete to harden before inspection
**Q133. Which precaution must be taken when working with powder-actuated tools?**
A. Train operators and verify material suitability before firing
B. Use excessive loads to ensure penetration
C. Allow multiple workers to share one tool without instruction
D. Point the tool toward a solid surface for testing
**Q134. What hazard is created when materials are stored unevenly on elevated platforms?**
A. Unbalanced weight leading to tipping or collapse
B. Faster material retrieval
C. Improved accessibility
D. Increased structural load capacity
**Q135. Which control best minimizes electrocution risk during wet weather?**
A. Operating barefoot to reduce grounding
B. Using GFCI-protected circuits and insulated tools
C. Delaying grounding verification until dry
D. Wearing leather gloves
**Q136. What is the main purpose of conducting hazard communication training?**
A. To comply with recordkeeping requirements
B. To ensure workers understand chemical risks and control measures
C. To reduce the need for PPE
D. To replace supervision duties
**Q137. Which measure prevents static buildup in flammable liquid storage areas?**
A. Keeping containers sealed and grounded
B. Using plastic drums only
C. Allowing spills to evaporate naturally
D. Increasing ventilation speed
**Q138. When is a trench considered a confined space?**
A. When its depth exceeds 10 feet

B. When entry or exit is limited and hazardous atmospheres may develop
C. When water is visible at the bottom
D. When only one worker is present

**Q139. What engineering control reduces worker vibration exposure from jackhammers?**
A. Rotating tools between shifts only
B. Applying more downward force
C. Wearing thicker gloves
D. Using vibration-damped handles and maintaining equipment

**Q140. What should be done if cracks are noticed in a scaffold support member?**
A. Tag scaffold as unsafe and remove from service immediately
B. Reinforce cracks with temporary braces
C. Continue work but limit load
D. Mark cracks for future inspection

**Q141. Which control reduces inhalation risk from fiberglass insulation fibers?**
A. Wearing long-sleeve cotton shirts only
B. Using respirators, gloves, and long sleeves during installation
C. Applying compressed air to clean surfaces
D. Sweeping debris manually after work

**Q142. Which hazard arises from poor ventilation in enclosed painting operations?**
A. Increased drying time only
B. Accumulation of flammable vapors and toxic fumes
C. Uneven paint texture
D. Moisture buildup

**Q143. What procedure must be followed before lifting operations using rigging equipment?**
A. Inspect slings, hooks, and shackles for defects and certification
B. Perform lifts first to test strength
C. Select random rigging from storage
D. Skip inspection if previously used

**Q144. Which hazard is created when combustible materials are stored near welding stations?**
A. Heat buildup only
B. Fire ignition from sparks and spatter
C. Excessive tool wear
D. Low visibility

**Q145. What is the primary reason for using hearing protection in high-noise environments?**
A. To comply with PPE policy
B. To prevent permanent hearing loss and improve focus
C. To block background noise only
D. To reduce communication needs

**Q146. Which practice ensures safety during confined space entry when using gas monitors?**
A. Calibrate monitors daily and continuously record readings
B. Test once before entry only
C. Place monitor outside the confined space
D. Ignore low-level alarm signals

**Q147. Which measure minimizes risk from hand-arm vibration syndrome (HAVS)?**
A. Using tool rest periods, anti-vibration gloves, and ergonomic scheduling
B. Continuous tool operation to build endurance
C. Applying additional grip pressure
D. Ignoring early tingling symptoms

**Q148. What step prevents falling object hazards when working under crane loads?**
A. Wearing helmets only
B. Establishing exclusion zones and preventing personnel entry beneath loads
C. Allowing access for observation
D. Using smaller cranes
**Q149. What control prevents exposure to harmful dust when cutting gypsum boards?**
A. Dry cutting with high-speed saws
B. Using dust extraction systems and respiratory protection
C. Blowing dust away using compressed air
D. Cutting in enclosed, unventilated rooms
**Q150. What safety measure should be taken when working near open excavations during night shifts?**
A. Use reflective barriers, adequate lighting, and secure fencing
B. Rely solely on warning signs
C. Allow vehicle parking near edges
D. Minimize lighting to avoid distraction
**Q151. Which document defines specific roles, responsibilities, and procedures to follow during an on-site emergency?**
A. Safety Data Sheet (SDS)
B. Job Hazard Analysis (JHA)
C. Emergency Action Plan (EAP)
D. Risk Assessment Form
**Q152. When conducting a post-incident investigation, which factor is most critical to determine first?**
A. The cost of damages
B. The sequence of events
C. The disciplinary actions needed
D. The insurance claim procedure
**Q153. What is the main purpose of conducting emergency drills on construction sites?**
A. To test workers' physical endurance
B. To evaluate emergency response effectiveness
C. To reduce insurance premiums
D. To meet daily training requirements
**Q154. Which type of fire extinguisher is suitable for use on electrical fires?**
A. Class A
B. Class B
C. Class C
D. Class D
**Q155. The Incident Command System (ICS) structure is primarily designed to:**
A. Control cost accounting during projects
B. Establish a clear chain of command during emergencies
C. Replace safety committees
D. Supervise construction productivity
**Q156. During an emergency evacuation, what is the designated area where personnel assemble for a headcount?**
A. Control Room
B. Muster Point
C. Safety Office
D. Staging Area
**Q157. Which method best identifies root causes in accident investigations?**

A. Five Whys analysis
B. Cost-benefit analysis
C. SWOT analysis
D. Statistical sampling

**Q158. In the event of a chemical spill, the first response priority is to:**
A. Notify the supplier
B. Evacuate affected personnel to a safe area
C. Begin containment immediately
D. File an incident report

**Q159. Which of the following is NOT typically included in an emergency action plan?**
A. Fire evacuation routes
B. Employee training schedule
C. Alarm systems and procedures
D. Reporting methods for emergencies

**Q160. Which person within the ICS coordinates communication among different responding agencies?**
A. Liaison Officer
B. Safety Officer
C. Operations Chief
D. Logistics Chief

**Q161. The primary goal of an incident investigation is to:**
A. Assign blame to responsible individuals
B. Prevent recurrence by identifying root causes
C. Document injury costs for insurance
D. Protect company reputation

**Q162. When developing an emergency response plan, which step should be completed first?**
A. Conducting hazard identification and risk assessment
B. Establishing media communication guidelines
C. Procuring emergency supplies
D. Scheduling annual drills

**Q163. Which term describes an incident that did not cause injury but had the potential to do so?**
A. Lost-time event
B. Recordable accident
C. Near miss
D. Unsafe act

**Q164. Which emergency management phase focuses on restoring normal operations after an incident?**
A. Response
B. Recovery
C. Mitigation
D. Preparedness

**Q165. During an evacuation, what is the supervisor's first responsibility?**
A. Retrieve safety documentation
B. Ensure all personnel are accounted for
C. Secure company equipment
D. Notify the media

**Q166. What is the correct sequence for managing an emergency incident?**
A. Respond → Recover → Mitigate → Prepare
B. Mitigate → Prepare → Respond → Recover
C. Prepare → Respond → Mitigate → Recover

D. Prepare → Recover → Respond → Mitigate
**Q167. Which factor most influences the effectiveness of emergency communication systems?**
A. The brand of radio used
B. Size of the facility
C. Color coding of equipment
D. Clarity and redundancy in communication channels
**Q168. What is the primary purpose of conducting a root cause analysis after an incident?**
A. To satisfy regulatory requirements
B. To determine preventive and corrective actions
C. To improve disciplinary consistency
D. To identify witness credibility
**Q169. Which type of emergency response training ensures workers can handle equipment like extinguishers and spill kits effectively?**
A. Theoretical lectures
B. Tabletop exercises
C. Hands-on practical training
D. Written assessments
**Q170. In fire prevention, maintaining clear exit routes primarily supports which principle?**
A. Safe access and egress
B. Confined space control
C. PPE utilization
D. Energy isolation
**Q171. When investigating an incident, what should be the investigator's first action upon arriving at the scene?**
A. Collect statements from witnesses
B. Secure and preserve the scene
C. Take photographs after cleanup
D. Review previous safety audits
**Q172. What does the term "universal precautions" refer to in emergency response?**
A. Treating all materials as hazardous waste
B. Handling all human bodily fluids as potentially infectious
C. Wearing full chemical-resistant suits at all times
D. Isolating every employee during medical emergencies
**Q173. Which key performance indicator best measures emergency preparedness?**
A. Number of drills conducted annually
B. Employee turnover rate
C. Maintenance cost of safety equipment
D. Number of inspections passed
**Q174. In emergency planning, which element ensures that different hazards are managed under one coordinated framework?**
A. Hazard-specific response plans
B. All-hazards approach
C. Task-specific SOPs
D. Contractor-specific safety plans
**Q175. Which factor most improves employee response during real emergencies?**
A. Fear of disciplinary consequences
B. Annual policy reviews by management
C. Relying solely on written procedures

D. Frequent and realistic simulation training

**Q176. Which step in emergency response focuses on identifying and reducing potential hazards before an incident occurs?**

A. Preparedness
B. Mitigation
C. Recovery
D. Response

**Q177. What is the main advantage of using the Incident Command System (ICS) in large-scale emergencies?**

A. It eliminates the need for communication between agencies
B. It limits authority to one organization
C. It provides a unified, flexible command structure
D. It replaces site-specific emergency plans

**Q178. When assessing emergency preparedness, what does the term "resilience" refer to?**

A. The ability to prevent hazards entirely
B. The ability to quickly recover and continue operations
C. The number of resources allocated for emergencies
D. The amount of funding in safety programs

**Q179. During an investigation, why is it important to interview witnesses as soon as possible?**

A. To control rumors
B. To capture accurate, fresh recollections
C. To finalize reports early
D. To complete compliance paperwork

**Q180. What is the first priority when responding to a workplace fire?**

A. Notify the insurance company
B. Record property damage
C. Investigate the fire source
D. Rescue endangered personnel safely

**Q181. What is the purpose of conducting a debriefing after an emergency drill?**

A. To penalize poor performers
B. To identify improvement opportunities and lessons learned
C. To revise the training schedule
D. To reduce drill frequency

**Q182. Which control measure best limits injury from falling objects during an emergency evacuation in a construction area?**

A. Wearing high-visibility vests
B. Establishing controlled egress routes
C. Increasing site lighting
D. Using signage at exits

**Q183. In incident reporting, what is the difference between a hazard and a risk?**

A. Hazards are conditions; risks are likelihood and severity outcomes
B. Hazards are outcomes; risks are conditions
C. They are interchangeable
D. Hazards always have zero probability

**Q184. When a crane contacts a live power line, what is the safest immediate action for the operator?**

A. Exit the cab immediately
B. Stay inside until power is confirmed off
C. Jump out and run

D. Call for help from the boom

**Q185. Which emergency phase includes implementing corrective measures based on previous incident evaluations?**

A. Preparedness
B. Recovery
C. Mitigation
D. Response

**Q186. In fire safety management, why should combustible waste be removed regularly?**

A. To maintain a clean appearance
B. To reduce fuel sources that increase fire intensity
C. To comply with environmental policies
D. To provide space for storage

**Q187. Which activity occurs during the preparedness phase of emergency management?**

A. Reviewing lessons learned post-event
B. Conducting training and drills
C. Assessing infrastructure damage
D. Removing hazardous materials

**Q188. What is the most important reason to keep accurate records of emergency drills?**

A. To demonstrate compliance and evaluate performance trends
B. To determine payroll adjustments
C. To share results with the public
D. To avoid re-training requirements

**Q189. What factor most commonly leads to communication breakdown during emergencies?**

A. Use of too many communication channels
B. Ambiguous or conflicting instructions
C. Lack of proper uniforms
D. Excessive planning documentation

**Q190. During an emergency, the safety officer's main duty is to:**

A. Approve procurement requests
B. Direct evacuation logistics
C. Supervise financial transactions
D. Monitor responder health and safety

**Q191. In an incident involving hazardous gas release, what immediate action should be taken after alerting personnel?**

A. Begin evacuation upwind from the release
B. Attempt to plug the leak
C. Call maintenance to inspect
D. Record the time in the logbook

**Q192. Which best describes the role of an emergency coordinator?**

A. Maintain equipment inventory
B. Direct and control on-site emergency response activities
C. Conduct monthly safety audits
D. Supervise procurement

**Q193. What is the first step in developing an emergency response plan for a construction project?**

A. Identify likely emergency scenarios
B. Train all subcontractors
C. Post evacuation maps
D. Purchase response equipment

**Q194. What is the most effective way to prevent panic during emergencies?**
A. Conducting frequent realistic drills
B. Displaying safety posters
C. Appointing new supervisors
D. Installing additional alarms
**Q195. Which of the following best describes the recovery phase's main goal?**
A. Conducting new hazard analysis
B. Restoring operations and infrastructure to normal
C. Initiating rescue operations
D. Launching root cause investigations
**Q196. When performing an accident investigation, why is collecting physical evidence important?**
A. It satisfies recordkeeping policies only
B. It replaces witness testimony
C. It supports factual analysis and prevents speculation
D. It allows quicker closure of the report
**Q197. During emergency response training, what element ensures participants retain learned procedures?**
A. Hands-on simulation and repetition
B. Reading lengthy manuals
C. Passive observation
D. Annual written tests only
**Q198. Which control system ensures accountability of personnel entering and exiting hazardous zones?**
A. Buddy system
B. Permit-to-work system
C. Access card record
D. Shift scheduling
**Q199. In a confined space emergency, what must rescuers always do before entering?**
A. Enter quickly to save time
B. Test the atmosphere and obtain authorization
C. Open all access points immediately
D. Deactivate communication systems
**Q200. What is the purpose of a business continuity plan in emergency management?**
A. To manage long-term recovery and operational resilience
B. To coordinate volunteer activities
C. To guide law enforcement investigations
D. To detail employee compensation

---

## Answer keys and Explanations (101-200)

**Q101. Answer: D**

**Explanation**: The safest approach is coordinating with the utility company to de-energize or relocate lines before any work begins. De-energization removes the hazard entirely rather than relying solely on PPE or warning systems. Signs and tape provide awareness but cannot prevent electrocution from accidental contact or arcing near energized lines.

**Q102. Answer: D**

**Explanation**: Chemical-resistant PPE provides barrier protection, while eyewash stations enable immediate neutralization in case of splashes. Open-air conditions or frequent rinsing do not provide adequate prevention or emergency response. Proper PPE and emergency infrastructure mitigate both exposure and damage severity.

**Q103. Answer: D**

**Explanation**: LOTO ensures hazardous energy sources are isolated and tagged, preventing accidental activation that could harm workers. It is a preventive safety protocol, not a productivity or convenience measure. Guards and other devices remain essential but do not replace LOTO's protective function.

**Q104. Answer: A**

**Explanation**: Insulated, moisture-wicking fabrics maintain body temperature and dryness, while warm-up breaks restore circulation. Cotton absorbs sweat and increases cooling risk. Sustained exposure without rest rapidly leads to cold injuries and reduced worker performance.

**Q105. Answer: D**

**Explanation**: Insufficient lighting obscures visual cues, making uneven steps or obstacles difficult to detect, leading to trip or impact injuries. Noise and ventilation remain unaffected. Regular lighting checks ensure safe passage and clear visibility during both day and night operations.

**Q106. Answer: B**

**Explanation**: Exhaust ventilation removes harmful gases, while catalytic converters lower emission toxicity. Worker rotation manages exposure time but doesn't eliminate contaminants. Odor masking disguises rather than mitigates the hazard. Engineering control ensures air quality and compliance with occupational exposure limits.

**Q107. Answer: A**

**Explanation**: Engineering assessments determine load distribution and safe demolition sequences, ensuring adjacent structure stability. Uncontrolled demolition or speed prioritization increases collapse risk. Structured sequencing provides predictability, maintaining site integrity and worker safety.

**Q108. Answer: D**

**Explanation**: UV-resistant PPE and welding screens absorb or block radiation, protecting skin and eyes. Distance and sunlight offer no protection against arc UV intensity. Full shielding ensures workers and bystanders avoid burns and eye injuries such as welder's flash.

**Q109. Answer: A**

**Explanation**: Keeping the load upgrade prevents tipping by maintaining center of gravity alignment, and slow speeds enhance stability. Turning sideways or overloading destabilizes the forklift. Proper orientation and speed minimize the chance of overturns and load shifts.

**Q110. Answer: B**

**Explanation**: Stored hydraulic energy can cause sudden fluid release and serious injury. Fully depressurizing and locking out systems prevents unintentional activation. Tightening or covering parts does not address residual pressure hazards effectively.

**Q111. Answer: C**

**Explanation**: Wet removal prevents fiber release, and negative pressure keeps contaminants confined. Dry sweeping and general airflow spread fibers, increasing exposure risk. Enclosed, controlled removal is the most effective method for preventing airborne contamination.

**Q112. Answer: B**
**Explanation**: Overextended planks act as levers, increasing bending stress and tipping potential. This leads to sudden breakage or loss of balance. Correct overhang limits ensure load distribution and platform stability during use, preventing structural failure.
**Q113. Answer: C**
**Explanation**: Approved containers prevent leakage, and upright positioning avoids vapor buildup. Mixing or open storage promotes ignition or contamination. Securing loads ensures vehicle stability and reduces spill hazards during transit, maintaining both fire safety and compliance.
**Q114. Answer: C**
**Explanation**: Mechanical and team lifting distribute weight evenly, minimizing strain on individuals. Rushing or ignoring warm-up increases injury likelihood. Proper ergonomics and shared handling are fundamental in preventing back and muscle injuries during manual tasks.
**Q115. Answer: D**
**Explanation**: Limiting air pressure to under 30 psi and using chip guards prevents debris injection injuries. High pressure or poor PPE usage can cause severe lacerations or blindness. Controlling both pressure and direction ensures cleaning safety and regulatory compliance.
**Q116. Answer: D**
**Explanation**: High-visibility clothing ensures detection, while eye contact with operators confirms mutual awareness. Proximity and distractions like phones increase risk. Direct communication and visible apparel prevent collisions and enhance coordination around mobile machinery.
**Q117. Answer: B**
**Explanation**: Inspection confirms integrity, and anchorage ensures force is transferred safely. Damaged or unsuitable equipment fails during falls. Using body belts or movable attachments violates safety standards, compromising effectiveness and increasing injury severity.
**Q118. Answer: A**
**Explanation**: Exceeding capacity alters balance and stress distribution, leading to tip-overs or component failure. Performance benefits are negligible and dangerous. Adhering to rated limits preserves stability and operational safety throughout lift usage.
**Q119. Answer: D**
**Explanation**: Bonding and grounding equalize electrical potential, preventing static sparks that could ignite vapors. Plastic containers and ungrounded operations increase ignition risk. Proper bonding is crucial for all transfers, regardless of duration or liquid type.
**Q120. Answer: D**
**Explanation**: Proper fall protection and secure anchor points prevent fatal falls during high-level inspections. Unprotected climbing or leaning creates unnecessary hazards. Using safety harnesses and anchored lines enables thorough inspection without compromising safety.
**Q121. Answer: A**
**Explanation**: Spotters provide external visual guidance where blind spots or nearby workers exist. Limiting usage to specific conditions overlooks common hazards. Spotter coordination prevents struck-by incidents and enhances control around congested work areas.
**Q122. Answer: C**
**Explanation**: Regular inspection ensures rigging integrity, and safety latches prevent load disengagement. Fast lifting and unsafe positioning create unpredictable movements. Equipment verification and exclusion zones guarantee stable, safe hoisting every time.
**Q123. Answer: D**
**Explanation**: Damaged cords expose conductors, and ungrounded systems fail to redirect current. Water enhances conductivity, drastically increasing electrocution risk. Grounding, double insulation, and GFCIs are essential to maintain safety during wet operations.

**Q124. Answer: B**
**Explanation**: Placing ladders near active doors can cause collision when the door opens unexpectedly. Blocking or securing doors and posting warnings prevent impact. Ladder angle or structural issues are unrelated unless combined with poor placement practices.
**Q125. Answer: D**
**Explanation**: Wet cutting suppresses dust generation, and vacuum collection removes residual particles at the source. Cloth masks lack filtration efficiency, and positioning downwind only shifts exposure. Engineering controls provide reliable silica dust mitigation during cutting tasks.
**Q126. Answer: A**
**Explanation**: High winds destabilize scaffolds and materials, so securing equipment and postponing work during unsafe conditions prevents falls or collapses. Adding personnel or removing guardrails increases hazard exposure. Proper judgment and weather monitoring maintain both structure and worker stability.
**Q127. Answer: D**
**Explanation**: Uncured concrete's alkaline compounds cause chemical burns upon prolonged skin contact. Thermal or electrical hazards are unrelated unless secondary conditions exist. Waterproof gloves and protective clothing prevent alkali exposure and long-term skin damage.
**Q128. Answer: B**
**Explanation**: Barricades restrict access, and tag lines stabilize swinging loads, preventing falls and struck-by injuries. Verbal instructions alone are inadequate under noisy conditions. Combining physical controls with communication protocols ensures edge safety during hoisting.
**Q129. Answer: C**
**Explanation**: Reviewing the SDS provides crucial information on hazards, PPE, and emergency procedures. Improper storage or testing poses serious health risks. Worker training ensures correct handling and compliance with chemical safety requirements under OSHA's Hazard Communication Standard.
**Q130. Answer: A**
**Explanation**: Coordination between multiple workers maintains stability and spatial control, preventing impacts with obstacles or walls. Vertical or dragging transport causes imbalance and property damage. Pre-clearing paths ensures smooth and safe movement through tight areas.
**Q131. Answer: D**
**Explanation**: Heat exposure increases internal pressure, risking cylinder rupture or valve failure. Shade and ventilation prevent over-pressurization. Corrosion occurs over time but is secondary to the immediate risk of explosive release caused by temperature elevation.
**Q132. Answer: B**
**Explanation**: Pre-pour inspection ensures formwork can withstand weight and pressure. Continuous monitoring detects deformation or shifting early. Pouring too quickly or using untrained labor increases collapse risk. Proper engineering verification ensures stable concrete containment.
**Q133. Answer: A**
**Explanation**: Trained operators verify both the tool and surface compatibility, preventing ricochet and material failure. Random testing or overloading creates uncontrolled discharge risks. Proper certification and site-specific evaluation ensure controlled, safe use.
**Q134. Answer: A**
**Explanation**: Uneven loads shift the platform's center of gravity, risking tipping or partial collapse. Balanced placement and securing materials prevent accidents. Convenience or speed should never compromise load distribution integrity and worker safety.
**Q135. Answer: B**
**Explanation**: GFCIs interrupt current flow instantly upon detecting fault, and insulated tools prevent contact transmission. Leather gloves absorb water and lose resistance. Reliable electrical safety requires pre-use checks and appropriate equipment during moisture exposure.

**Q136. Answer: B**
**Explanation**: Training equips workers with knowledge to identify, handle, and respond to chemical hazards safely. Recordkeeping and compliance are secondary benefits. Understanding risks promotes proactive behavior and consistent adherence to safe handling protocols.
**Q137. Answer: A**
**Explanation**: Grounding neutralizes charge differences, preventing static sparks that could ignite vapors. Plastic containers accumulate charge, increasing fire potential. Proper sealing and bonding procedures eliminate electrostatic discharge hazards effectively.
**Q138. Answer: B**
**Explanation**: A trench becomes confined when limited access and potential for toxic or oxygen-deficient air exist. Depth alone does not determine confined classification. Monitoring atmospheric conditions ensures safety before and during entry operations.
**Q139. Answer: D**
**Explanation**: Vibration-damped handles reduce transmitted shock, and well-maintained tools minimize mechanical imbalance. Gloves offer limited damping, while excessive pressure increases strain. Combining maintenance and ergonomic equipment design ensures worker health preservation.
**Q140. Answer: A**
**Explanation**: Visible damage indicates structural compromise. Immediate tagging and removal prevent catastrophic collapse. Reinforcement or delay allows hazard escalation. Inspections by competent personnel determine necessary repairs before re-use.
**Q141. Answer: B**
**Explanation**: Respiratory and skin protection prevent fiber irritation and inhalation. Cotton-only coverage lacks filtration capability. Air blasting or sweeping disperses fibers further, increasing airborne exposure risks. PPE and gentle cleanup ensure respiratory protection.
**Q142. Answer: B**
**Explanation**: Lack of airflow causes buildup of volatile organic compounds, creating fire and poisoning hazards. Ventilation systems dilute and expel fumes safely. Surface imperfections or humidity are minor compared to explosion and health risks from vapor concentration.
**Q143. Answer: A**
**Explanation**: Regular inspection ensures rigging integrity and verifies load capacity compliance. Skipping checks risks equipment failure, while random selection ignores safety ratings. Documentation and pre-use inspection are critical to preventing rigging accidents.
**Q144. Answer: B**
**Explanation**: Sparks easily ignite nearby combustibles, leading to rapid fire spread. Heat accumulation and tool wear are minor by comparison. Maintaining clear zones and using fire-resistant barriers protect both workers and property from ignition sources.
**Q145. Answer: B**
**Explanation**: Consistent exposure to excessive noise damages auditory cells irreversibly. Hearing protection prevents such loss and improves concentration by reducing fatigue. Policy compliance is secondary to protecting worker health and communication effectiveness.
**Q146. Answer: A**
**Explanation**: Calibration guarantees accuracy, and continuous monitoring detects atmospheric changes in real time. Testing only once or disregarding alarms exposes workers to toxic or oxygen-deficient environments. Proper monitoring prevents fatalities during entry operations.
**Q147. Answer: A**
**Explanation**: Scheduled breaks and anti-vibration gloves reduce exposure time and transmitted force. Ignoring early symptoms leads to chronic nerve damage. Combining administrative and PPE controls protects workers from long-term musculoskeletal disorders.

**Q148. Answer: B**
**Explanation**: Exclusion zones eliminate exposure by keeping workers out of potential impact areas. Helmets reduce minor injuries but cannot stop high-energy impacts. Restricting access ensures safety even if rigging failure occurs unexpectedly.
**Q149. Answer: B**
**Explanation**: Dust extraction removes particulates at the source, and respirators protect workers directly. Dry cutting and air blasting worsen airborne dust. Proper ventilation and PPE maintain clear air and reduce respiratory irritation.
**Q150. Answer: A**
**Explanation**: Proper lighting improves visibility, reflective barriers highlight hazards, and fencing prevents falls or equipment intrusion. Signage alone lacks physical prevention, and reduced lighting conceals risks. Comprehensive visibility control ensures 24-hour site safety around open excavations.
**Q151. Answer: C**
**Explanation**: An Emergency Action Plan outlines the steps employees must take in response to workplace emergencies, detailing evacuation routes, reporting methods, and key personnel duties. It ensures coordinated response actions to minimize harm. Unlike a JHA or risk assessment, which identify risks, an EAP focuses on immediate response measures. SDSs contain chemical hazard data, not procedural emergency instructions.
**Q152. Answer: B**
**Explanation**: Establishing the event sequence helps investigators understand the root causes and contributing factors leading to an incident. This chronological reconstruction allows for accurate causal analysis and preventive action. Damage costs and administrative processes follow later. Without a clear sequence, the investigation risks focusing on outcomes instead of causes.
**Q153. Answer: B**
**Explanation**: Drills provide practical assessment of the effectiveness of emergency plans and team coordination. They identify weaknesses in communication, timing, and evacuation routes. Their intent is not physical testing or insurance-related performance but ensuring readiness for real emergencies. Routine drills also build confidence and familiarity among workers.
**Q154. Answer: C**
**Explanation**: Class C extinguishers are designed for electrical fires, typically using non-conductive agents like carbon dioxide or dry chemicals to suppress flames safely. Using water-based extinguishers could create electrical shock hazards. Class A and B are for ordinary combustibles and flammable liquids, while Class D is for combustible metals.
**Q155. Answer: B**
**Explanation**: ICS ensures an organized, hierarchical command during incidents, clarifying authority, communication, and operational roles. It standardizes multi-agency coordination, ensuring efficient decision-making. It is not for productivity or budgeting purposes. Safety committees may exist separately, focusing on policy rather than emergency response execution.
**Q156. Answer: B**
**Explanation**: The muster point is the pre-identified safe location where personnel assemble after evacuation to verify accountability. It allows supervisors to identify missing persons quickly and coordinate with rescue teams. The staging area is used for equipment and responder coordination, not for general personnel gathering.
**Q157. Answer: A**
**Explanation**: The Five Whys technique helps investigators trace immediate and underlying causes by repeatedly questioning why an event occurred. This structured approach goes beyond surface factors to reveal systemic weaknesses. Other methods like cost-benefit or SWOT serve management functions, not causal investigation.

**Q158. Answer: B**
**Explanation**: Protecting life is always the top priority in spill response. Evacuating affected individuals reduces exposure risks and allows responders to act safely. Containment follows once safety is secured. Reporting and supplier notifications occur after stabilization and control measures are complete.
**Q159. Answer: B**
**Explanation**: While training supports preparedness, specific schedules are not part of the EAP itself. The plan focuses on procedures, responsibilities, and communication during emergencies. Training plans are typically included in safety program documentation, not in the immediate operational EAP.
**Q160. Answer: A**
**Explanation**: The Liaison Officer manages inter-agency coordination, ensuring clear communication between the incident command and external organizations. This role prevents duplication of efforts and maintains situational awareness. The Safety Officer focuses on responder protection, while Operations and Logistics Chiefs handle field tasks and resource support respectively.
**Q161. Answer: B**
**Explanation**: Effective investigations aim to identify underlying causes and prevent future incidents through corrective measures. Focusing on blame undermines learning and safety improvement. Insurance and reputation management are secondary benefits but not the investigation's core intent.
**Q162. Answer: A**
**Explanation**: Hazard identification and risk assessment form the foundation for all response planning. Understanding what could go wrong determines resource allocation and response protocols. Without this analysis, plans may fail to address real threats effectively or allocate resources efficiently.
**Q163. Answer: C**
**Explanation**: A near miss is an unplanned event that could have resulted in injury, illness, or damage but did not. Recording and analyzing near misses help prevent future incidents by identifying overlooked hazards. Lost-time events involve actual injuries, while unsafe acts are behaviors that may lead to incidents.
**Q164. Answer: B**
**Explanation**: The recovery phase addresses restoration of operations, equipment, and morale following an emergency. It includes reviewing lessons learned and implementing improvements. Mitigation reduces future risks, preparedness ensures readiness, and response is the immediate action phase during the event.
**Q165. Answer: B**
**Explanation**: Supervisors must ensure all team members safely exit and assemble at the muster point. Accountability is essential for identifying missing persons quickly. Retrieving materials or communicating with external parties takes lower priority compared to ensuring personnel safety and headcount accuracy.
**Q166. Answer: B**
**Explanation**: The emergency management cycle begins with mitigation (reducing hazards), followed by preparedness (planning and training), response (immediate action), and recovery (returning to normal). This structured order ensures proactive control of risks before, during, and after incidents.
**Q167. Answer: D**
**Explanation**: Clear, redundant communication ensures consistent message delivery under stress. Systems must be fail-safe, using multiple channels (radios, alarms, verbal instructions) to avoid breakdowns. Equipment brand or facility size has lesser impact compared to clarity, redundancy, and training.
**Q168. Answer: B**
**Explanation**: Root cause analysis identifies fundamental process or system failures, guiding preventive and corrective actions. Its purpose goes beyond compliance—it drives long-term safety improvement. Discipline and witness evaluation may support the process but are not the analysis objective.
**Q169. Answer: C**

**Explanation**: Practical, hands-on training builds muscle memory and confidence in using response tools such as extinguishers and spill kits. It bridges the gap between theory and real action. Tabletop or written methods reinforce understanding but lack experiential realism necessary for critical performance.

**Q170. Answer: A**

**Explanation**: Clear exits ensure safe access and egress, enabling quick evacuation during emergencies. Obstructed pathways increase risk of injury or entrapment. While PPE and energy control contribute to safety, they do not directly address fire evacuation or escape efficiency.

**Q171. Answer: B**

**Explanation**: Preserving the scene prevents loss or alteration of evidence, allowing accurate reconstruction of events. Witness interviews and documentation follow once safety is confirmed. Cleanup before documentation risks destroying crucial clues that explain the incident's origin.

**Q172. Answer: B**

**Explanation**: Universal precautions assume all bodily fluids may contain infectious agents, prompting consistent protective measures like gloves, masks, and proper disposal. This standard safeguards responders regardless of confirmed infection. It differs from general chemical or isolation procedures.

**Q173. Answer: A**

**Explanation**: Frequent, well-documented drills demonstrate ongoing preparedness and allow evaluation of response efficiency. While inspections and maintenance support readiness, they do not directly measure response capability. Turnover rate reflects HR factors, not emergency readiness.

**Q174. Answer: B**

**Explanation**: The all-hazards approach integrates multiple emergency types—fire, chemical, weather—into one cohesive response system, improving coordination and flexibility. It avoids redundant, siloed plans and streamlines resource management. Hazard-specific and task-based plans may fit under this overarching framework.

**Q175. Answer: D**

**Explanation**: Realistic simulations reinforce learning under stress, improving retention and reaction accuracy during actual emergencies. Regular exposure reduces panic and confusion. Written policies and management reviews are supportive but less effective without hands-on, experiential reinforcement.

**Q176. Answer: B**

**Explanation**: Mitigation involves proactive efforts to reduce or eliminate hazards before emergencies occur. It includes engineering controls, process changes, and preventive measures designed to minimize risk severity. Preparedness focuses on training and planning, while response and recovery happen after the event. Mitigation ensures fewer emergencies and lowers their impact.

**Q177. Answer: C**

**Explanation**: ICS allows diverse agencies to coordinate effectively under a common structure. It ensures clear communication, chain of command, and adaptable resource management. Unlike rigid hierarchies, ICS can expand or contract to match incident complexity. It supplements, not replaces, site-specific emergency procedures.

**Q178. Answer: B**

**Explanation**: Resilience measures an organization's capacity to absorb disruption and resume critical operations swiftly after emergencies. It combines effective planning, adaptability, and learning from past events. Prevention focuses on hazard elimination, while resilience focuses on recovery and continuity. High resilience reduces downtime and long-term impacts.

**Q179. Answer: B**

**Explanation**: Immediate interviews capture reliable, detailed accounts before memory fades or perceptions change. Quick communication also prevents misinformation from spreading. Delay risks memory distortion, reducing the accuracy of event reconstruction. Early witness engagement supports stronger factual conclusions and corrective action planning.

**Q180. Answer: D**
**Explanation**: Human safety is the primary focus in any fire emergency. Ensuring that all personnel evacuate and receive aid overrides all other actions. Property and documentation concerns follow only after all people are accounted for. This life-first principle underlies all emergency response standards and ethics.
**Q181. Answer: B**
**Explanation**: Post-drill debriefs analyze performance, identify communication or procedural weaknesses, and recommend enhancements. They convert simulated experiences into learning opportunities for continuous improvement. The goal is constructive evaluation, not punishment, ensuring that each drill strengthens readiness for future real incidents.
**Q182. Answer: B**
**Explanation**: Designated egress routes minimize movement under active work zones or unstable structures, reducing risk from falling debris. Controlled routes are vital in dynamic environments like construction sites. Visibility and lighting enhance awareness but cannot physically protect workers from hazards above.
**Q183. Answer: A**
**Explanation**: A hazard represents a potential source of harm, such as a chemical or unstable structure. Risk expresses the likelihood and severity of injury from that hazard. Understanding both concepts allows accurate prioritization of corrective actions. Mislabeling them leads to poor hazard control strategies.
**Q184. Answer: B**
**Explanation**: The safest option is to remain in the cab because exiting can cause electrocution due to ground potential differences. Operators should wait until power is de-energized and clearance is confirmed. Only if fire threatens should they jump clear with feet together to avoid current flow through the body.
**Q185. Answer: C**
**Explanation**: Mitigation uses data and lessons from past incidents to strengthen future safety. It includes redesigning processes, enhancing engineering controls, and revising plans to reduce future impact. Preparedness focuses on readiness; recovery restores operations; response deals with active emergencies.
**Q186. Answer: B**
**Explanation**: Combustible waste provides fuel that can accelerate or intensify fires. Routine removal minimizes ignition potential and helps maintain safe, unobstructed workspaces. A clean site also improves visibility and evacuation efficiency, contributing to comprehensive fire prevention practices.
**Q187. Answer: B**
**Explanation**: Preparedness includes developing plans, conducting training, and running drills to ensure readiness for emergencies. It builds individual and organizational capabilities for effective response. Damage assessment and hazard removal occur in recovery and mitigation phases respectively, not preparedness.
**Q188. Answer: A**
**Explanation**: Accurate drill records verify compliance with regulations and help track performance improvement over time. Documentation supports audits and reveals training gaps. Sharing with the public or payroll-related reasons do not serve the main safety management purpose of continual improvement.
**Q189. Answer: B**
**Explanation**: Confusing or contradictory directions cause hesitation and errors during crisis communication. Clear, concise, and consistent messaging is vital to maintain order. Multiple channels can improve redundancy, but clarity remains the key to effectiveness under pressure.
**Q190. Answer: D**
**Explanation**: The Safety Officer ensures responder safety by monitoring conditions, identifying hazards, and enforcing protective measures. This role advises the incident commander on risks during response operations. Other logistics and financial functions belong to different ICS roles.

**Q191. Answer: A**
**Explanation**: Evacuation upwind ensures personnel move away from the contaminated area, minimizing exposure. Containment efforts should only be performed by trained responders with proper PPE. Logging and maintenance calls are secondary to immediate life-saving actions.
**Q192. Answer: B**
**Explanation**: The emergency coordinator leads on-site response efforts, ensures coordination with external agencies, and verifies execution of the emergency plan. They act as the focal point for communication and control, ensuring procedures align with organizational policy and local regulations.
**Q193. Answer: A**
**Explanation**: Scenario identification defines what types of emergencies could occur, guiding the rest of the planning process. Without this foundational analysis, training and resources may not match actual risks. It ensures focus on credible, site-specific threats like fire, collapse, or chemical exposure.
**Q194. Answer: A**
**Explanation**: Regular, realistic drills build familiarity and confidence, reducing fear and confusion during real incidents. Practice helps workers trust procedures and respond calmly. Posters and alarms support awareness but do not replace behavioral conditioning through hands-on rehearsal.
**Q195. Answer: B**
**Explanation**: Recovery emphasizes restoring normal function, including operations, equipment, and personnel well-being. It transitions the organization from emergency mode back to steady-state operations. Rescue and analysis activities occur earlier, during response or post-incident phases.
**Q196. Answer: C**
**Explanation**: Physical evidence anchors findings in objective facts, validating witness statements and identifying technical causes. Tangible data prevent bias and speculation, leading to credible corrective actions. Sole reliance on verbal accounts can result in incomplete or inaccurate conclusions.
**Q197. Answer: A**
**Explanation**: Active participation through simulation enhances memory retention and confidence under stress. Repetition reinforces behavioral consistency, improving performance during real events. Passive learning and reading provide theoretical knowledge but lack the experiential reinforcement required for practical recall.
**Q198. Answer: B**
**Explanation**: The permit-to-work system controls and documents entry into hazardous or restricted areas, ensuring only authorized and trained personnel access them. It maintains accountability and enforces pre-entry safety checks. The buddy system complements but doesn't formally record such movements.
**Q199. Answer: B**
**Explanation**: Atmospheric testing ensures oxygen and toxin levels are safe before entry, and authorization confirms that a rescue plan and PPE are in place. Immediate, unplanned entry often results in additional casualties. Safe rescue protocols depend on pre-entry hazard assessment.
**Q200. Answer: A**
**Explanation**: Business continuity planning ensures critical operations continue or resume quickly after disruptions. It integrates with emergency response and recovery to maintain essential services, protecting productivity and stakeholder confidence. It is not about law enforcement or financial compensation processes.

## Questions (201-300)

**Q201. Which document should be used by construction workers to evaluate hazards before starting a job?**

A. Emergency Action Plan (EAP)
B. Job Hazard Analysis (JHA)
C. Incident Report Form
D. Risk Management Plan

**Q202. Which of the following is the primary responsibility of the Incident Commander during an emergency?**

A. Supervising the media response
B. Ensuring the safety of all responders and victims
C. Coordinating logistics and supplies
D. Monitoring employee attendance

**Q203. What is the first step in responding to a chemical spill?**

A. Notify external authorities
B. Evacuate all personnel from the area
C. Assess the situation and identify the material
D. Begin containment procedures

**Q204. During an emergency evacuation, which of the following should be prioritized?**

A. Saving company property
B. Ensuring all employees are accounted for
C. Completing safety documentation
D. Making sure all equipment is turned off

**Q205. What is the purpose of conducting a post-incident review in emergency management?**

A. To assess employee performance and assign blame
B. To identify areas for improvement and update response plans
C. To review financial losses
D. To determine employee eligibility for compensation

**Q206. In emergency response, which of the following is considered the most effective means of communicating with the public?**

A. Social media updates
B. Press releases
C. Telephone calls
D. Public address systems

**Q207. What is the primary objective of an emergency response team during a workplace fire?**

A. To rescue victims from a collapsed building
B. To control and extinguish the fire
C. To protect valuable equipment
D. To gather witness statements

**Q208. Which of the following best describes an emergency action plan (EAP)?**

A. A document outlining long-term disaster recovery
B. A list of safety equipment for all workers
C. A strategy for responding to specific emergency situations
D. A database of emergency contacts

**Q209. How can supervisors prevent injury during emergency evacuations?**

A. By ensuring proper PPE is worn
B. By controlling crowd movement and ensuring calmness

C. By monitoring equipment shutdowns
D. By performing hazard analysis after the evacuation

**Q210. What is the correct procedure for managing a medical emergency at the workplace?**
A. Start CPR immediately
B. Evacuate all personnel before offering assistance
C. Wait for the victim to recover naturally
D. Call emergency services and provide first aid as needed

**Q211. Which term is used to describe injuries or illnesses caused by prolonged exposure to a hazard at work?**
A. Acute injury
B. Chronic illness
C. Immediate harm
D. Permanent injury

**Q212. What should be included in a fire safety training program for employees?**
A. Use of fire extinguishers and evacuation procedures
B. Details of insurance claims processes
C. The location of all employee offices
D. A list of all hazardous materials used

**Q213. In an emergency involving multiple casualties, which approach ensures effective treatment?**
A. Triage to prioritize patients based on severity
B. Treating patients on a first-come, first-served basis
C. Focusing only on the most severely injured
D. Waiting for external medical teams to arrive

**Q214. Which of the following best describes the role of an emergency response coordinator?**
A. Conduct safety audits regularly
B. Design and implement safety programs
C. Oversee emergency response training and drills
D. Manage company finances during a crisis

**Q215. During a hazardous material spill, what action should be taken to ensure personnel safety?**
A. Evacuate all non-essential personnel and secure the area
B. Attempt to clean the spill immediately
C. Wait for emergency services to arrive and do nothing
D. Continue work as normal, ensuring minimal disruption

**Q216. Which of the following best describes a near miss?**
A. An incident that causes severe injury
B. An event where an injury or damage was narrowly avoided
C. An injury that occurs without any external cause
D. A workplace injury that results in lost time

**Q217. In the context of emergency management, what does the term "continuity" refer to?**
A. Maintaining a normal business environment during and after a crisis
B. The uninterrupted flow of work
C. Continuous operation of a single department
D. The speed of recovery from an incident

**Q218. How can an organization improve emergency response effectiveness?**
A. By increasing staff numbers during non-emergency times
B. By conducting regular drills and reviewing lessons learned
C. By cutting costs on emergency supplies
D. By avoiding involvement in external emergency training programs

**Q219. What is the purpose of using a fire watch during hot work activities?**
A. To monitor for signs of fire and act quickly if one starts
B. To ensure compliance with environmental laws
C. To keep track of workers' hours
D. To monitor air quality during the work process
**Q220. What is the purpose of implementing a "permit-to-work" system?**
A. To ensure all workers are certified before starting
B. To document hazardous work and ensure proper controls are in place
C. To monitor employee performance
D. To assign supervisors to work areas
**Q221. Which emergency management phase includes evaluating and analyzing recovery efforts?**
A. Response
B. Mitigation
C. Preparedness
D. Recovery
**Q222. What is the role of a safety officer during an emergency?**
A. To lead search and rescue operations
B. To ensure safety and well-being of all responders and personnel
C. To oversee financial aspects of the crisis
D. To manage the communications with media
**Q223. Which type of emergency response equipment is most crucial for confined space rescues?**
A. Fire extinguishers
B. Breathing apparatus and retrieval systems
C. Chemical spill kits
D. Personal protective equipment
**Q224. What should be the first priority for emergency responders at a hazardous material spill?**
A. Contain the material immediately
B. Evacuate surrounding areas
C. Identify the material and its risks
D. Begin cleanup procedures
**Q225. Which of the following is NOT an element of the Incident Command System (ICS)?**
A. Chain of command
B. Resource allocation
C. Employee performance evaluations
D. Communication structure
**Q226. Which of the following is a key element of a business continuity plan?**
A. Detailed employee contact information
B. Procedures for maintaining critical operations during a disaster
C. The inventory of all company property
D. A list of potential media outlets for crisis communication
**Q227. In an emergency involving a large crowd, what should responders prioritize first?**
A. Securing the area and preventing unauthorized access
B. Evacuating people as quickly as possible
C. Ensuring the safety of emergency responders
D. Establishing clear communication with the public
**Q228. What is the best method to ensure that all employees are familiar with emergency evacuation routes?**
A. Distribute written evacuation plans

B. Post evacuation plans in common areas
C. Conduct regular evacuation drills
D. Email the evacuation plans to all employees

**Q229. Which emergency response role is responsible for managing logistics and resource allocation?**
A. Incident Commander
B. Safety Officer
C. Operations Chief
D. Logistics Chief

**Q230. In the event of an active shooter on-site, which of the following is the most appropriate response?**
A. Wait for law enforcement to arrive before evacuating
B. Try to disarm the shooter if possible
C. Evacuate and secure a safe area as quickly as possible
D. Attempt to negotiate with the shooter

**Q231. What is the primary purpose of a site-specific safety plan (SSSP)?**
A. To outline disciplinary actions for unsafe behavior
B. To detail emergency response procedures only
C. To define tailored safety measures addressing unique project risks
D. To replace the company's overall safety manual

**Q232. Which document forms the foundation of an effective construction safety management system?**
A. Organizational chart
B. Safety policy statement
C. Toolbox meeting agenda
D. Material inventory list

**Q233. What is the most effective way to measure the success of a safety program implementation?**
A. Counting PPE distributed
B. Tracking reductions in incident rates and severity trends
C. Evaluating training attendance
D. Monitoring audit completion percentages

**Q234. What should be included in a construction company's safety policy to ensure accountability?**
A. Generic safety slogans
B. Optional reporting procedures
C. Broad encouragement statements
D. Defined responsibilities and authority levels for each role

**Q235. Which best describes a leading indicator in safety performance measurement?**
A. A recordable injury rate
B. Results of near-miss reporting and safety audits
C. Workers' compensation claim history
D. Total number of lost workdays

**Q236. Which action most strengthens worker participation in a safety program?**
A. Mandating attendance without feedback opportunities
B. Encouraging hazard reporting and including workers in decision-making
C. Limiting communication to supervisors
D. Reducing worker involvement to maintain efficiency

**Q237. Which element is crucial when conducting a safety audit?**
A. Evaluating compliance with regulatory and internal standards
B. Reviewing project schedules only
C. Comparing worker productivity levels
D. Interviewing senior management exclusively

**Q238. How should corrective actions from safety inspections be prioritized?**
A. Based on cost of implementation
B. According to worker preference
C. By evaluating potential severity and likelihood of harm
D. In alphabetical order of identified hazards
**Q239. What is the key outcome of implementing a safety incentive program correctly?**
A. Reduced reporting of injuries
B. Genuine improvement in safe behaviors and hazard identification
C. Increased paperwork for supervisors
D. Lower productivity due to additional oversight
**Q240. What should be verified during safety program audits to confirm documentation accuracy?**
A. Frequency of file updates
B. Consistency between field practices and written procedures
C. Volume of stored documents
D. Number of sign-offs per page
**Q241. Which step ensures new regulatory changes are integrated into an existing safety program?**
A. Review by competent personnel followed by policy updates and retraining
B. Ignoring updates unless mandated by audit
C. Waiting for annual review cycles
D. Applying them only to new projects
**Q242. What does a thorough accident trend analysis help an organization accomplish?**
A. Delay audits
B. Justify disciplinary actions
C. Increase recordable incidents
D. Identify recurring hazards and system weaknesses
**Q243. Why is management commitment critical to a safety program's success?**
A. It provides funding, leadership, and visible example for compliance
B. It reduces the need for worker participation
C. It allows delegation of all responsibilities
D. It ensures minimal legal exposure only
**Q244. Which component ensures consistency across multiple construction project safety programs?**
A. Centralized corporate safety framework adaptable to each site
B. Independent plans without coordination
C. Informal communication among safety officers
D. Varying procedures per supervisor preference
**Q245. Which documentation practice best supports continuous safety improvement?**
A. Storing completed reports without review
B. Reviewing, analyzing, and feeding findings into program updates
C. Discarding old audit results
D. Delegating documentation solely to interns
**Q246. Which safety management element ensures systematic identification and mitigation of hazards?**
A. Periodic random checks
B. Risk assessment and control planning
C. Emergency drill frequency
D. Management performance reviews
**Q247. What is the purpose of establishing a safety committee on a large project?**
A. To assign blame for incidents
B. To facilitate communication and collaboration between management and workers

C. To replace supervisor responsibilities
D. To manage payroll and attendance

**Q248. What primary objective does job hazard analysis (JHA) serve within a safety program?**
A. Documenting hazards for recordkeeping only
B. Assigning blame post-incident
C. Identifying job-specific hazards and implementing control measures
D. Meeting training attendance requirements

**Q249. Which practice most effectively ensures sustainability of a safety program?**
A. One-time training sessions
B. Continuous monitoring, feedback, and adaptive updates
C. Rigid procedures with no revisions
D. Limiting employee involvement

**Q250. Which document typically outlines the employer's responsibilities in a construction safety program?**
A. Project schedule
B. Safety management plan
C. Personnel evaluation form
D. Tool inventory list

**Q251. Which element must be verified during a management safety review?**
A. Current performance against defined objectives and indicators
B. Individual attendance records only
C. Payroll adjustments for safety staff
D. Daily weather logs

**Q252. What is the main reason for maintaining accurate safety training records?**
A. To increase administrative workload
B. To ensure proof of compliance and verify worker competency
C. To reduce field supervision
D. To simplify payroll documentation

**Q253. Why should safety procedures be reviewed after a significant incident?**
A. To determine if existing controls failed and require revision
B. To punish responsible personnel
C. To document insurance claims
D. To minimize management involvement

**Q254. Which process ensures new employees understand their safety responsibilities before beginning work?**
A. Providing manuals without discussion
B. On-the-job mentoring without instruction
C. Signing forms without explanation
D. Orientation and induction training covering hazards and expectations

**Q255. What is the key benefit of integrating safety performance into overall project metrics?**
A. It aligns safety with productivity and quality objectives for holistic success
B. It isolates safety from other project functions
C. It focuses solely on compliance statistics
D. It minimizes emphasis on operational outcomes

**Q256. What is the primary function of a safety audit in a construction program?**
A. To assign blame for accidents
B. To identify gaps in compliance and improve safety systems
C. To reduce the number of inspections

D. To replace risk assessments

**Q257. Which step ensures the credibility of data collected during safety inspections?**

A. Conducting inspections by qualified personnel using standardized checklists
B. Allowing each supervisor to design their own checklist
C. Recording only major hazards
D. Ignoring environmental conditions during observation

**Q258. What defines a proactive safety culture in construction organizations?**

A. Responding quickly after incidents
B. Encouraging hazard reporting and addressing risks before accidents occur
C. Waiting for audits to identify deficiencies
D. Minimizing documentation to save time

**Q259. Which approach best sustains long-term safety performance improvement?**

A. Short-term campaigns only
B. Integrating safety goals into business objectives and performance appraisals
C. Relying solely on incident investigations
D. Conducting annual training without feedback

**Q260. Why is worker feedback essential during safety program evaluations?**

A. It provides real-world insights into effectiveness of controls and procedures
B. It replaces management inspections entirely
C. It focuses primarily on morale issues
D. It limits program modifications

**Q261. Which step ensures effective implementation of corrective actions following a safety audit?**

A. Assigning responsibility, deadlines, and follow-up verification
B. Filing the report without action
C. Waiting for external inspectors to verify improvements
D. Prioritizing low-cost corrections only

**Q262. What is the benefit of benchmarking safety performance against industry standards?**

A. It eliminates need for audits
B. It provides promotional content for marketing
C. It reveals competitive advantage and areas for improvement
D. It replaces internal evaluations

**Q263. Why are pre-task safety meetings critical in daily construction operations?**

A. They review specific hazards and controls before each task begins
B. They replace formal safety training
C. They serve as attendance tracking opportunities
D. They satisfy documentation requirements only

**Q264. Which management behavior most influences strong safety culture development?**

A. Delegating all responsibilities to safety officers
B. Leading by example through consistent safety participation
C. Avoiding involvement in toolbox meetings
D. Focusing on production targets exclusively

**Q265. Which method most effectively evaluates the impact of safety training programs?**

A. Observing post-training behavior changes and reduced incident rates
B. Counting attendance alone
C. Relying on post-course tests only
D. Ignoring performance outcomes

**Q266. What is the purpose of a management of change (MOC) process in safety systems?**

A. To assess risks before introducing new equipment, processes, or materials

B. To document staffing changes only
C. To simplify purchasing procedures
D. To replace hazard assessments

**Q267. Which action demonstrates continuous improvement within a safety program?**
A. Ignoring feedback from incident investigations
B. Maintaining unchanged procedures over time
C. Reviewing performance metrics regularly and updating goals accordingly
D. Reducing program scope to cut costs

**Q268. Which tool helps evaluate compliance and procedural adherence objectively?**
A. Safety audit checklist and inspection forms
B. Verbal reports from workers
C. Informal discussions with managers
D. Random interviews only

**Q269. How should senior management respond to recurring safety deficiencies identified in audits?**
A. Conduct root cause analysis and allocate resources for corrective measures
B. Ignore them if incidents are minor
C. Delegate correction solely to field workers
D. Wait for external consultants to intervene

**Q270. Why is integration of environmental management within safety programs beneficial?**
A. It separates responsibilities further
B. It adds unnecessary complexity
C. It focuses solely on waste disposal
D. It promotes holistic risk management across health, safety, and environmental domains

**Q271. What is a common pitfall when implementing behavior-based safety (BBS) programs?**
A. Blaming individuals rather than addressing system deficiencies
B. Observing behaviors regularly
C. Reinforcing positive actions
D. Using anonymous observation data

**Q272. What role does documentation play in maintaining ISO 45001 or similar system certifications?**
A. Provides traceability and evidence of compliance with defined processes
B. Serves as marketing collateral
C. Replaces physical inspections
D. Eliminates need for training

**Q273. Which control ensures subcontractor alignment with project safety requirements?**
A. Allowing independent safety practices
B. Awarding contracts to the lowest bidder only
C. Prequalification based on safety performance and program compatibility
D. Ignoring past safety records

**Q274. Why should companies maintain a written safety program manual?**
A. It serves as a central reference outlining policies, roles, and procedures
B. It replaces practical supervision
C. It fulfills a filing requirement only
D. It limits flexibility in operations

**Q275. What is a key outcome of incorporating technology, such as mobile inspection apps, into safety management?**
A. Real-time data collection and faster hazard resolution
B. Increased paperwork burden
C. Reduced worker engagement

D. Elimination of audits

**Q276. What is essential for effective corrective action tracking after a safety audit?**

A. Filing reports without follow-up
B. Assigning responsibility, setting deadlines, and verifying closure with evidence
C. Waiting for annual reviews
D. Delegating tracking to temporary staff

**Q277. Which practice best ensures subcontractor safety alignment on complex projects?**

A. Contracting based solely on lowest bid
B. Prequalification, ongoing monitoring, and contractual safety obligations
C. Allowing subcontractors to follow their own rules
D. Ignoring subcontractor safety history

**Q278. What is the primary purpose of a management of change (MOC) procedure in a safety program?**

A. To document staffing changes only
B. To replace job hazard analyses
C. To speed up procurement processes
D. To assess and control risks before implementing new equipment, materials, or procedures

**Q279. How should organizations evaluate the effectiveness of safety training over time?**

A. By counting attendees only
B. By measuring on-the-job behavior change, assessments, and incident trends related to trained tasks
C. By issuing certificates only
D. By limiting training to managers

**Q280. What is the critical focus of a high-quality incident investigation?**

A. Assigning immediate blame to individuals
B. Determining root causes, identifying systemic failures, and implementing corrective actions with verification
C. Only documenting the event for records
D. Closing the investigation quickly to resume work

**Q281. What is the first step in developing an effective construction safety program?**

A. Conducting a comprehensive hazard assessment of the work environment
B. Scheduling employee orientations
C. Creating incident reports
D. Purchasing personal protective equipment

**Q282. Which management activity most strengthens accountability for safety performance?**

A. Conducting annual audits only
B. Delegating safety to only the safety department
C. Including safety goals in performance evaluations for all personnel
D. Outsourcing all monitoring to consultants

**Q283. What is a key indicator of effective communication within a safety management system?**

A. Workers demonstrate understanding of hazards and reporting procedures
B. Management issues lengthy policy statements
C. Posters and slogans are displayed prominently
D. Emails are sent regularly without follow-up

**Q284. Which principle underlies an effective health and safety management system?**

A. Plan–Do–Check–Act (PDCA) continuous improvement cycle
B. Fixed one-time policy creation
C. Annual inspections without data analysis
D. Delegating responsibility solely to line supervisors

**Q285. What component ensures an organization's safety program aligns with applicable laws and standards?**
A. Peer reviews without legal oversight
B. Periodic regulatory reviews by competent professionals
C. Supervisor discretion
D. Random safety talks

**Q286. Which element differentiates a strong safety culture from a compliant one?**
A. Voluntary reporting and peer accountability beyond minimum regulations
B. Meeting inspection checklists only
C. Maintaining silence after incidents
D. Relying exclusively on external audits

**Q287. Why should lessons learned from incident investigations be communicated across all departments?**
A. To prevent recurrence and promote organizational learning beyond the affected area
B. To assign blame publicly
C. To satisfy documentation requirements
D. To increase disciplinary records

**Q288. What distinguishes an internal safety audit from a regulatory inspection?**
A. Internal audits proactively identify gaps for improvement before enforcement involvement
B. Internal audits replace government oversight completely
C. Regulatory inspections are voluntary
D. Internal audits focus only on paperwork

**Q289. Which best practice enhances employee retention within effective safety programs?**
A. Limiting feedback to management only
B. Increasing workload to compensate for compliance costs
C. Recognizing safe performance and involving workers in decision-making
D. Reducing training time for efficiency

**Q290. What ensures that safety procedures remain relevant and effective over time?**
A. Periodic review, employee feedback, and updates based on new technologies and incidents
B. Keeping original procedures unchanged
C. Outsourcing all documentation maintenance
D. Removing worker input from revisions

**Q291. What role does management leadership play in fostering a safety culture?**
A. Demonstrating commitment through visible involvement and resource allocation
B. Delegating authority without participation
C. Communicating only after incidents occur
D. Reducing safety budgets to save costs

**Q292. Why is it important to evaluate contractor safety performance periodically?**
A. To delay payment until evaluation ends
B. To verify continued compliance and ensure consistent hazard control effectiveness
C. To transfer risk entirely to the contractor
D. To reduce documentation requirements

**Q293. Which method best ensures proper implementation of emergency response plans?**
A. Conducting regular drills and debriefs involving all relevant personnel
B. Waiting until an incident occurs
C. Limiting participation to supervisors
D. Storing plans without training

**Q294. What is the most effective approach for addressing recurring minor incidents?**

A. Conducting root cause analysis and implementing systemic corrective actions
B. Ignoring them since they are minor
C. Focusing only on recordable cases
D. Blaming individual workers

**Q295. Which feature enhances the sustainability of a construction safety program?**
A. Integration into core business processes and leadership accountability
B. Restricting safety to compliance departments
C. One-time safety campaigns
D. Annual inspections only

**Q296. Which key factor determines the quality of an organization's risk assessments?**
A. Performing assessments annually only
B. Conducting assessments informally
C. Relying solely on generic templates
D. Competence and multidisciplinary participation of assessors

**Q297. How can companies ensure effective communication of revised safety procedures?**
A. Providing structured briefings, updated manuals, and acknowledgment sign-offs
B. Assuming workers will learn informally
C. Posting changes without explanation
D. Informing only supervisors

**Q298. Which monitoring method provides the most accurate insight into daily safety practices?**
A. Unannounced field observations and behavioral audits
B. Reviewing reports quarterly only
C. Waiting for external inspections
D. Discussing performance once per year

**Q299. What is a critical step in integrating ergonomics into construction safety management?**
A. Replacing manual handling with verbal warnings
B. Waiting until injury reports increase
C. Assessing physical demands and adapting tasks, tools, or workstations
D. Ignoring worker feedback

**Q300. Why should management review contractor safety documentation before project mobilization?**
A. To verify plan adequacy and alignment with site-specific safety expectations
B. To delay project start for administrative purposes
C. To transfer all safety responsibility to contractors
D. To complete recordkeeping requirements

---

## Answer keys and Explanations (201-300)

**Q201. Answer: B**

**Explanation**: Job Hazard Analysis (JHA) identifies and assesses hazards related to specific tasks. It ensures workers understand risks and controls before work begins. The EAP focuses on emergencies, while incident forms and risk plans help after hazards are identified but are not used for prior task assessments.

**Q202. Answer: B**

**Explanation**: The Incident Commander's primary duty is to manage the overall safety and effectiveness of the emergency response. This includes ensuring the safety of both responders and victims. Coordination, logistics, and media management fall under the responsibility of other ICS roles.

**Q203. Answer: C**

**Explanation**: The first step is assessing the situation, which includes identifying the spilled material, its hazards, and its potential impact. This ensures the proper containment, evacuation, and notification protocols are followed. Attempting containment without proper identification could lead to further hazards.

**Q204. Answer: B**

**Explanation**: The priority during evacuation is ensuring all employees are accounted for and safely evacuated. Securing property, equipment, and completing paperwork can occur later. Accountability is critical for preventing loss and ensuring no one is left behind.

**Q205. Answer: B**

**Explanation**: A post-incident review identifies weaknesses in response strategies and uncovers areas for improvement. This analysis helps refine plans, enhance training, and avoid recurrence. Assigning blame or focusing solely on financial aspects detracts from the goal of improving future response.

**Q206. Answer: A**

**Explanation**: Social media provides the fastest and most direct communication with the public during emergencies. It allows for real-time updates, quick distribution of critical information, and addressing concerns. Traditional methods like press releases and PA systems are useful but slower in comparison.

**Q207. Answer: B**

**Explanation**: The main role of an emergency response team in a fire is to control and extinguish the fire as quickly as possible to prevent further damage. Saving equipment or gathering statements comes after the fire is under control, and rescuing victims occurs as part of the overall safety effort.

**Q208. Answer: C**

**Explanation**: An Emergency Action Plan (EAP) outlines the procedures to follow during specific emergency scenarios, such as evacuations, fire response, or medical emergencies. It is a key tool in ensuring a coordinated and effective response. It is not focused on recovery or contact management.

**Q209. Answer: B**

**Explanation**: Controlling crowd movement and maintaining calmness helps prevent panic, confusion, and injury during evacuations. Supervisors should focus on leading and directing personnel to designated safe areas while ensuring that all exits remain clear and accessible. PPE and hazard analysis play supporting roles but are secondary.

**Q210. Answer: D**

**Explanation**: The first step is to contact emergency services for professional medical help. Simultaneously, provide basic first aid, such as CPR or wound care, if necessary. Immediate intervention can save lives. Waiting or evacuating first without assisting would delay vital care.

**Q211. Answer: B**

**Explanation**: Chronic illness is the result of prolonged exposure to hazardous conditions, such as repetitive strain, toxic chemicals, or physical stress. It develops gradually over time, unlike acute injuries, which occur suddenly. Long-term exposure management is critical in preventing such illnesses.

**Q212. Answer: A**

**Explanation**: A fire safety training program should cover practical skills like using fire extinguishers and understanding evacuation routes. It also includes knowledge of fire prevention and safety systems. Insurance claims and office locations are irrelevant to immediate safety actions during a fire emergency.

**Q213. Answer: A**

**Explanation**: Triage is used to prioritize medical treatment based on injury severity, ensuring that the most critical patients are treated first. This system prevents overwhelming healthcare resources and maximizes life-saving potential. Delaying or treating based on arrival time or severity alone is ineffective in mass casualty scenarios.

**Q214. Answer: C**

**Explanation**: The emergency response coordinator manages emergency preparedness, ensuring effective response during crises. This includes overseeing drills, training, and readiness exercises. Financial management and safety audits are separate responsibilities within the organization.

**Q215. Answer: A**

**Explanation**: The first priority is evacuating non-essential personnel and securing the area to prevent exposure. Handling hazardous materials without proper training or PPE increases risks. Immediate containment should only be attempted by trained responders with the correct equipment.

**Q216. Answer: B**

**Explanation**: A near miss is an event where damage or injury was narrowly avoided but could have occurred. Reporting near misses is crucial for identifying hazards before they result in real harm. Unlike actual injuries, near misses help prevent future accidents by addressing latent risks.

**Q217. Answer: A**

**Explanation**: Continuity refers to the ability of an organization to maintain essential operations during and after an emergency. Business continuity planning includes preparations for disruptions and ensures quick restoration of critical services, minimizing the overall impact of a crisis.

**Q218. Answer: B**

**Explanation**: Regular drills and post-event reviews improve the organization's ability to respond effectively. These activities help identify weaknesses, enhance coordination, and ensure that lessons from past emergencies are integrated into future plans. Cutting supplies or avoiding external training limits preparedness.

**Q219. Answer: A**

**Explanation**: A fire watch is essential during hot work activities to spot potential fire hazards, particularly after work is completed. They are trained to act quickly if a fire starts, reducing the risk of uncontrolled fires. Environmental compliance and air quality monitoring are separate responsibilities.

**Q220. Answer: B**

**Explanation**: A permit-to-work system is designed to ensure that all hazardous activities are properly planned, controlled, and monitored before they begin. It includes specific safety precautions, personnel qualifications, and equipment checks to prevent accidents. It is not used for tracking performance or supervisor assignments.

**Q221. Answer: D**

**Explanation**: Recovery is the phase where efforts to restore normalcy are evaluated. It involves assessing the effectiveness of the response, making adjustments, and applying lessons learned to improve future preparedness. Mitigation and preparedness occur before and during the incident, while response is the immediate action phase.

**Q222. Answer: B**

**Explanation**: The Safety Officer's primary role is to ensure the safety of responders and all personnel involved in the emergency response. They assess risks, ensure compliance with safety procedures, and monitor hazardous conditions. Other roles such as search and rescue or media communication fall under different ICS positions.

**Q223. Answer: B**
**Explanation**: Breathing apparatus and retrieval systems are critical in confined space rescues due to the risk of low oxygen levels or toxic atmospheres. These tools ensure rescuers can operate safely.
**Q224. Answer: C**
**Explanation**: The first step is to identify the spilled material and its associated risks. Understanding its properties and hazards allows responders to make informed decisions about containment, evacuation, and protective measures. Immediate containment or cleanup without identification may worsen the situation.
**Q225. Answer: C**
**Explanation**: The ICS is designed for efficient emergency management and includes a clear chain of command, resource allocation, and communication structures to coordinate efforts. Employee performance evaluations are not part of the ICS; this system focuses on managing the response to an incident.
**Q226. Answer: B**
**Explanation**: A business continuity plan ensures that critical business functions can continue or resume quickly after a disruption. It includes strategies to maintain operations, such as backup systems, alternate work locations, and resource management. Employee contact information and media lists are supplementary but not central to the plan.
**Q227. Answer: B**
**Explanation**: The first priority is evacuating people to safety. Responders should focus on guiding people away from danger zones, maintaining order, and preventing panic. Securing the area and communication are important but secondary to immediate life-saving actions.
**Q228. Answer: C**
**Explanation**: Regular evacuation drills help employees practice using the routes and become familiar with them under pressure. While posting or distributing plans supports awareness, hands-on experience through drills is essential to ensure employees can navigate effectively during an actual emergency.
**Q229. Answer: D**
**Explanation**: The Logistics Chief is responsible for managing the resources required for emergency response, including supplies, equipment, and personnel. The Incident Commander oversees the overall response, while the Safety Officer focuses on the safety of all involved, and the Operations Chief handles field operations.
**Q230. Answer: C**
**Explanation**: Evacuating the area and securing a safe location is the most critical response in an active shooter situation. The priority is to ensure personal safety and avoid confrontation. Law enforcement should be the ones to handle the shooter. Trying to disarm or negotiate could escalate the danger.
**Q231. Answer: C**
**Explanation**: A site-specific safety plan customizes safety protocols for the particular hazards and conditions of a given project. Unlike general manuals, it integrates risk assessments, task analyses, and emergency measures suited to that job. This targeted approach ensures that workers face known hazards with proper controls, aligning safety planning with actual site operations.
**Q232. Answer: B**
**Explanation**: The safety policy statement expresses management's commitment and provides the strategic direction for safety performance. It communicates goals, responsibilities, and accountability. While charts and meetings support operations, the policy establishes the governing principle that guides all subsequent procedures and ensures alignment with regulatory compliance and company culture.
**Q233. Answer: B**
**Explanation**: Incident rate reductions and severity trends reflect true performance outcomes, showing how preventive measures impact worker safety. Tracking attendance or audits assesses activity, not

effectiveness. Continuous analysis of leading and lagging indicators verifies sustained improvement and program maturity.

**Q234. Answer: D**

**Explanation**: Clearly assigning safety responsibilities ensures each individual knows their duties and authority in enforcing safety. General encouragement lacks actionable structure. Defined accountability integrates safety into job functions, promoting ownership and measurable compliance across all organization levels.

**Q235. Answer: B**

**Explanation**: Leading indicators reflect proactive activities—such as inspections, audits, and near-miss reporting—that predict future outcomes. Lagging indicators like injury rates describe past performance. Tracking leading metrics allows early correction and continuous improvement before incidents occur.

**Q236. Answer: B**

**Explanation**: Inclusion empowers employees to contribute ideas and identify risks early, fostering a culture of ownership. Mandates without engagement create compliance without commitment. Open dialogue between management and field workers enhances awareness and accountability at every level.

**Q237. Answer: A**

**Explanation**: Audits assess conformance to safety regulations and internal controls, identifying deficiencies and opportunities for improvement. Examining only productivity or management limits scope. Comprehensive compliance evaluation ensures continuous alignment with safety requirements and best practices.

**Q238. Answer: C**

**Explanation**: Prioritization must focus on risk assessment—severity and probability guide which issues demand immediate attention. Cost or preference fails to address actual hazard significance. Risk-based ranking ensures that critical threats are controlled promptly for maximum impact.

**Q239. Answer: B**

**Explanation**: Effective incentives promote proactive safety behavior, such as hazard spotting and reporting, rather than concealing incidents. Mismanaged programs can distort reporting accuracy. Balanced recognition systems reinforce positive actions and sustainable behavioral change.

**Q240. Answer: B**

**Explanation**: True verification ensures that documented procedures align with real-world execution. Consistency confirms that safety policies are active and effective, not merely administrative. Overemphasis on file quantity overlooks operational validity and behavior alignment.

**Q241. Answer: A**

**Explanation**: Prompt evaluation and integration ensure ongoing compliance and consistent safety performance. Delayed or selective adoption leads to outdated practices. Continuous review and retraining embed new requirements into daily operations efficiently.

**Q242. Answer: D**

**Explanation**: Trend analysis reveals repeating issues and process gaps that cause recurring incidents. Recognizing these trends allows targeted corrective measures. Using data to improve systems is far more effective than punitive responses that suppress reporting.

**Q243. Answer: A**

**Explanation**: Leadership defines tone and priority; funding ensures resources for training and equipment. Workers emulate management's visible commitment, making safety integral to operations. True engagement builds trust and drives continuous improvement beyond compliance.

**Q244. Answer: A**

**Explanation**: A unified framework defines minimum standards and expectations, ensuring that each project maintains core safety elements while adapting to local risks. Disconnected programs cause inconsistency and confusion, undermining overall safety culture coherence.

**Q245. Answer: B**
**Explanation**: Continuous improvement depends on analyzing past performance to adjust procedures and policies. Static storage prevents learning. Regular reviews transform documentation into actionable intelligence, closing feedback loops and promoting proactive risk management.
**Q246. Answer: B**
**Explanation**: Risk assessment processes identify potential hazards and assign control measures systematically. Random checks lack structure, while reviews or drills target different objectives. Structured evaluation ensures preventive control before incidents occur.
**Q247. Answer: B**
**Explanation**: Safety committees bridge management and workforce communication, ensuring concerns are heard and corrective actions prioritized. They foster participation and ownership across job levels. Their role supports—not replaces—supervisory duties, creating a cooperative safety culture.
**Q248. Answer: C**
**Explanation**: A JHA systematically evaluates each task to identify associated hazards and establish preventive controls. It's a proactive tool aimed at prevention rather than documentation. Integrating JHAs ensures workers understand safe job execution before performing tasks.
**Q249. Answer: B**
**Explanation**: Sustainable safety programs evolve through continuous improvement cycles based on observation and data feedback. Static systems stagnate and fail to address new risks. Adaptation ensures ongoing relevance and compliance under changing conditions.
**Q250. Answer: B**
**Explanation**: The safety management plan defines management duties, communication flow, and accountability. It outlines employer commitment to hazard control, training, and enforcement. Other documents manage logistics or personnel but not safety structure and policy direction.
**Q251. Answer: A**
**Explanation**: Management reviews compare performance against established goals to evaluate program success. Reviewing unrelated administrative records provides no insight into safety outcomes. Systematic evaluation drives improvement and accountability through data-driven management oversight.
**Q252. Answer: B**
**Explanation**: Training records confirm regulatory compliance and worker readiness for assigned tasks. They also identify retraining needs and demonstrate due diligence during audits. Record accuracy supports continuous professional development and program credibility.
**Q253. Answer: A**
**Explanation**: Incident reviews identify procedural weaknesses and inform improvements to prevent recurrence. Assigning blame doesn't address root causes. Updating controls ensures lessons learned translate into strengthened safety management systems.
**Q254. Answer: D**
**Explanation**: Orientation sessions educate new hires about site hazards, reporting methods, and their duties. This builds foundational awareness essential for safe behavior. Passive paperwork or observation alone cannot substitute structured onboarding training.
**Q255. Answer: A**
**Explanation**: Including safety as a core project metric ensures it receives equal weight with cost and schedule goals. This integration fosters shared responsibility across departments and balances productivity with protection. Safety becomes a measurable, integral component of project success.
**Q256. Answer: B**
**Explanation**: A safety audit provides an objective review of program effectiveness, identifying weaknesses and opportunities for improvement. It compares actual practices with policies and regulatory requirements. The findings inform corrective actions, strengthen procedures, and verify system integrity.

Rather than punitive evaluation, audits serve as diagnostic tools promoting continuous improvement, compliance assurance, and safety culture maturity across the organization.

**Q257. Answer: A**

**Explanation**: Inspections performed by trained personnel using standardized tools ensure consistency, accuracy, and comparability across sites. This structure reduces bias and promotes data reliability. Ad hoc checklists cause inconsistency, and selective reporting misses critical information. Standardization enables meaningful trend analysis and enhances confidence in decisions based on collected data.

**Q258. Answer: B**

**Explanation**: A proactive culture focuses on early hazard detection, employee participation, and preventive action rather than reaction. Workers are empowered to report concerns without fear of reprisal. This continuous vigilance and improvement mindset transforms safety from compliance-driven to engagement-based, significantly reducing incident potential and fostering shared accountability.

**Q259. Answer: B**

**Explanation**: Embedding safety within business performance systems links accountability and continuous improvement directly to company success. This alignment motivates all levels to prioritize safe practices daily. Short-term or isolated actions produce temporary effects. Integration ensures consistent focus, measurable outcomes, and strategic alignment between operational efficiency and worker protection.

**Q260. Answer: A**

**Explanation**: Workers interact with controls daily, offering firsthand observations on practicality, clarity, and compliance barriers. Their input reveals gaps unseen in management reviews and supports realistic program adaptation. Incorporating worker perspectives promotes shared responsibility, reinforces trust, and improves the program's functional value beyond mere compliance.

**Q261. Answer: A**

**Explanation**: Assigning responsibility ensures accountability, deadlines establish urgency, and follow-up confirms closure. Without verification, deficiencies persist unnoticed. A structured corrective action process maintains audit integrity and demonstrates management's commitment to continuous improvement, ensuring findings lead to tangible safety enhancements rather than administrative formality.

**Q262. Answer: C**

**Explanation**: Benchmarking compares organizational results with recognized standards or peers to identify performance gaps and best practices. This encourages improvement through measurable targets and informed decision-making. Unlike promotional use, it serves as a diagnostic tool to assess alignment with regulatory and professional benchmarks for sustained progress.

**Q263. Answer: A**

**Explanation**: Pre-task meetings refresh hazard awareness, clarify safe work methods, and promote open communication. Addressing task-specific risks daily ensures adaptive control to changing conditions. They complement training by reinforcing knowledge and engaging crews directly in proactive hazard prevention and job readiness.

**Q264. Answer: B**

**Explanation**: Management's visible involvement in safety activities demonstrates authentic commitment, motivating workers to follow suit. Leadership by example sets the tone for workplace behavior. Passive delegation diminishes credibility and signals that safety is secondary, weakening program integrity and trust.

**Q265. Answer: A**

**Explanation**: Measuring how workers apply learned skills in real tasks and correlating it with improved safety metrics validates training effectiveness. Attendance or testing alone measures participation, not competency. Long-term behavioral and incident data confirm whether training translates into tangible field improvements.

**Q266. Answer: A**

**Explanation**: MOC identifies and controls risks associated with operational or procedural modifications. Evaluating potential impacts prevents unintended hazards from new conditions. Without this step, even minor changes can undermine established safety measures or introduce uncontrolled risks to personnel and property.

**Q267. Answer: C**

**Explanation**: Continuous improvement depends on data-driven reassessment and adaptive goal setting. Regular reviews align program objectives with evolving risks and performance trends. Static systems stagnate, while dynamic updates reinforce accountability, innovation, and proactive safety enhancement across all operations.

**Q268. Answer: A**

**Explanation**: Structured checklists and inspection forms provide quantifiable, consistent evaluation metrics, allowing accurate comparison across projects. Subjective discussions can highlight perceptions but lack verification. Standardized tools ensure transparency, measurable progress tracking, and compliance documentation integrity.

**Q269. Answer: A**

**Explanation**: Root cause analysis uncovers systemic issues behind repeated findings, while resource allocation ensures solutions are implemented effectively. Ignoring problems erodes credibility and compliance. Active management response demonstrates leadership engagement and long-term risk mitigation commitment.

**Q270. Answer: D**

**Explanation**: Integrating environmental and safety systems reduces duplication, aligns regulatory compliance, and supports sustainability. Shared objectives streamline risk assessments, communication, and corrective actions. This synergy fosters efficient management of interrelated hazards impacting both human health and the environment.

**Q271. Answer: A**

**Explanation**: Effective BBS focuses on identifying barriers to safe behavior, not assigning blame. Targeting individuals undermines trust and discourages reporting. Programs succeed when observation outcomes guide systemic improvement and continuous reinforcement of positive behaviors across teams.

**Q272. Answer: A**

**Explanation**: Documentation validates that procedures are implemented and effective. It supports audit readiness and ensures continuity even during staff changes. Marketing or substitution uses miss the purpose of transparency and accountability required by international safety management standards.

**Q273. Answer: C**

**Explanation**: Prequalification confirms subcontractors meet project safety expectations and regulatory standards before work begins. Evaluating past performance ensures reliability and compatibility. Bypassing assessment compromises program integrity and increases potential for inconsistent compliance and risk exposure.

**Q274. Answer: A**

**Explanation**: A written manual documents the organization's safety framework, guiding consistent application across departments. It defines responsibilities, hazard controls, and emergency protocols. Far from being restrictive, it enhances clarity and accountability, ensuring uniform compliance and employee understanding.

**Q275. Answer: A**

**Explanation**: Digital tools enable instant reporting and analytics, allowing quicker corrective action and trend identification. This immediacy enhances communication and transparency. Rather than replacing human input, technology complements inspections, streamlining documentation and enabling proactive decision-making through timely, accurate data.

**Q276. Answer: B**
**Explanation**: An effective corrective action tracking system ensures each finding is assigned to a responsible owner, given a target completion date, and monitored through verification steps until closure. This formalizes accountability and prevents repeat issues by documenting progress, resource allocation, and validation evidence. It also creates an auditable trail for management reviews and regulatory inspections, enabling trend analysis and demonstrating sustained commitment to hazard elimination and continuous program improvement.
**Q277. Answer: B**
**Explanation**: Prequalification processes evaluate subcontractor safety records, training, and program compatibility to ensure alignment with project standards. Requiring documentation of past incident rates, safety certifications, and proof of effective safety management demonstrates capability. Ongoing monitoring through site inspections, toolbox talks, and performance reviews maintains accountability. Clear contractual obligations, communication channels, and corrective action clauses ensure subcontractors address deficiencies promptly, preserving overall project safety integrity and reducing multi-employer risk exposure.
**Q278. Answer: D**
**Explanation**: An MOC process requires formal review of proposed changes to equipment, materials, or procedures, assessing hazards and determining necessary controls before implementation. It includes stakeholder involvement, updated documentation, training for affected personnel, and verification of control effectiveness. By standardizing risk evaluation and approval steps, MOC prevents unanticipated hazards from being introduced into operations. Regular audits of MOC records ensure compliance and institutional learning for future modifications and safer project execution.
**Q279. Answer: B**
**Explanation**: Evaluating training effectiveness goes beyond attendance; it measures knowledge retention, behavior change, and practical application on the job. This is achieved through post-training assessments, observations during tasks, and monitoring incident or near-miss trends linked to trained activities. Feedback loops refine course content and delivery, while refresher sessions address identified deficiencies. Demonstrating measurable improvements in safe practices validates training investment and supports continuous improvement in competency and site risk reduction and regulatory compliance.
**Q280. Answer: B**
**Explanation**: Effective incident investigations determine root causes by separating proximate events from underlying system failures, using methods like root cause analysis or the 5 Whys. Investigators gather evidence, interview witnesses impartially, and preserve scene integrity while avoiding blame. Recommendations must address systemic fixes—procedural changes, training, or engineering controls—and include implementation timelines and verification. Transparent reporting and shared lessons learned prevent recurrence, reinforce accountability, and improve organizational resilience and safety culture and stakeholder engagement.
**Q281. Answer: A**
**Explanation**: An effective safety program starts with a thorough hazard assessment to identify potential risks and exposures inherent to the construction site. This foundation informs control measures, training requirements, and resource allocation. Without an accurate understanding of hazards, later components such as procedures and training may fail to address critical risks, leading to preventable incidents and noncompliance with regulatory safety standards and performance goals.
**Q282. Answer: C**
**Explanation**: Integrating safety objectives into employee performance appraisals ensures responsibility and motivation at every level. When management and field workers alike are evaluated on safety outcomes, commitment becomes consistent and measurable. Delegation or infrequent audits lack

reinforcement. Aligning accountability with professional evaluation embeds safety as a shared value, reinforcing proactive risk management and fostering continuous improvement across the organization.

**Q283. Answer: A**

**Explanation**: True communication effectiveness is reflected in worker comprehension and application of safety procedures. Clear messaging, feedback mechanisms, and open dialogue ensure consistent hazard awareness. Policy distribution without engagement or verification does not confirm understanding. Interactive communication builds trust, reinforces compliance, and enhances overall program effectiveness and operational safety performance metrics and accountability.

**Q284. Answer: A**

**Explanation**: The PDCA model ensures systematic planning, implementation, monitoring, and improvement of safety programs. It fosters adaptability and learning through periodic review. Static policies lack flexibility, while single annual checks miss evolving risks. PDCA embeds safety management into organizational culture, driving consistency, data-informed decisions, and sustained excellence in occupational health performance, operational resilience, and regulatory compliance.

**Q285. Answer: B**

**Explanation**: Regular evaluations by qualified personnel confirm that policies comply with current regulations and industry standards. Changes in laws or codes must be quickly reflected in company procedures. Informal or inconsistent approaches can result in violations. Scheduled reviews ensure that compliance obligations remain integrated into the organization's safety management system and support continuous legal and ethical conformity and organizational accountability.

**Q286. Answer: A**

**Explanation**: A strong culture thrives on intrinsic motivation and mutual accountability rather than mere rule adherence. Workers actively identify and address hazards while management supports transparency. Compliance-driven cultures focus on external enforcement, whereas proactive ones internalize safety as core value. Sustained engagement, open reporting, and shared learning define long-term organizational safety excellence and risk prevention strategies across operations.

**Q287. Answer: A**

**Explanation**: Sharing lessons learned extends awareness of root causes, ensuring that corrective actions are replicated across the organization. This prevents similar failures in other divisions and enhances systemic resilience. Keeping information localized limits effectiveness. Open communication fosters transparency, interdepartmental learning, and a culture focused on prevention rather than punishment and reactive corrective measures.

**Q288. Answer: A**

**Explanation**: Internal audits serve as proactive self-assessments to improve compliance and performance before external enforcement identifies deficiencies. They provide early corrective opportunities and encourage continuous improvement. Regulatory inspections verify compliance externally and may impose penalties. Combining both processes strengthens system resilience and ensures readiness and sustained excellence in occupational safety management outcomes.

**Q289. Answer: C**

**Explanation**: Recognition and inclusion foster loyalty and engagement, showing workers that their safety and input are valued. Participatory programs reduce turnover by creating trust and shared responsibility. Conversely, neglecting involvement or overburdening staff decreases morale. Active engagement drives sustainable workforce stability and improved operational performance, aligning safety outcomes with overall organizational success.

**Q290. Answer: A**

**Explanation**: Regularly revising safety procedures ensures adaptability to evolving risks, technologies, and lessons from incidents. Engaging employees adds practical insight, ensuring procedures are realistic. Outdated methods compromise safety and compliance. Continuous updates strengthen alignment between policy and practice, supporting dynamic hazard control, regulatory adherence, and operational excellence across diverse construction environments.

**Q291. Answer: A**

**Explanation**: Leadership engagement through consistent presence, communication, and support establishes credibility and trust. Allocating resources for training and equipment demonstrates genuine care. Passive delegation or reactive management weakens culture. Active leadership drives accountability and motivates employees to integrate safety into everyday decision-making, reinforcing preventive behavior and promoting operational excellence throughout the organization.

**Q292. Answer: B**

**Explanation**: Periodic contractor evaluations confirm adherence to project-specific safety standards, verifying that controls remain effective throughout work phases. This continuous oversight prevents complacency and identifies improvement needs early. Evaluations foster collaboration and accountability, ensuring safe performance and alignment with client and regulatory expectations across all subcontracted activities.

**Q293. Answer: A**

**Explanation**: Drills transform written plans into practiced responses, revealing weaknesses and building competence. Inclusive participation ensures everyone understands roles and coordination procedures. Documentation alone cannot guarantee preparedness. Regular rehearsals and reviews refine readiness, reduce panic during real events, and strengthen organizational resilience and confidence in handling emergencies effectively and safely.

**Q294. Answer: A**

**Explanation**: Even small incidents reveal latent system flaws that can lead to severe accidents if unaddressed. Root cause analysis uncovers underlying deficiencies and drives corrective measures. Ignoring or blaming prevents learning and perpetuates risk. Systemic evaluation fosters proactive prevention and continuous improvement in overall safety performance, culture, and management accountability.

**Q295. Answer: A**

**Explanation**: Embedding safety into business strategy ensures it remains a fundamental operational value, supported by leadership accountability and metrics. Standalone initiatives fade over time. Integration sustains focus and aligns safety outcomes with production, quality, and environmental performance, building long-term resilience and organizational reliability in project execution and workforce protection.

**Q296. Answer: D**

**Explanation**: Competent assessors with diverse expertise ensure comprehensive identification and evaluation of hazards from multiple perspectives. Generic or informal approaches miss unique risks. Effective assessments combine technical knowledge, practical site experience, and stakeholder input, producing accurate risk controls and ensuring alignment with operational realities and safety objectives for continuous improvement.

**Q297. Answer: A**

**Explanation**: Structured briefings and accessible updates guarantee comprehension and confirm accountability through acknowledgment. Simply posting or relying on informal methods creates gaps in understanding. Comprehensive rollout ensures every affected employee knows changes, why they matter, and how they apply, maintaining consistency and compliance throughout operations and team coordination.

**Q298. Answer: A**

**Explanation**: Spontaneous site observations capture real behaviors under normal conditions, providing genuine insight into compliance and cultural adherence. Regular audits identify improvement opportunities promptly. Infrequent reviews miss emerging issues. Continuous monitoring embeds accountability and supports immediate corrective coaching, reinforcing proactive safe work habits across all field operations.

**Q299. Answer: C**

**Explanation**: Ergonomic assessments identify repetitive strain and overexertion hazards, leading to tailored interventions such as tool redesign or lifting aids. Proactive adaptation prevents musculoskeletal disorders and improves productivity. Ignoring early signs leads to long-term injury costs. Integration supports worker health, project efficiency, and morale while reducing compensation claims and downtime.

**Q300. Answer: A**

**Explanation**: Reviewing contractor documentation ensures safety plans meet site hazards, regulatory standards, and client requirements before work begins. Early verification prevents inconsistencies that could lead to unsafe practices or project delays. This due diligence reinforces collaboration, sets expectations, and establishes accountability throughout construction phases for consistent safety performance and compliance.

## Questions (301-400)

**Q301. What distinguishes an effective safety management meeting from routine gatherings?**

A. Minimal management participation
B. Unstructured discussion of unrelated topics
C. Attendance tracking only
D. Focused review of performance indicators, incidents, and improvement actions

**Q302. Which control ensures consistent communication of company safety expectations to all workers?**

A. Conducting regular toolbox talks covering job-specific hazards
B. Distributing policies once annually
C. Posting rules without discussion
D. Allowing supervisors to interpret independently

**Q303. What should be reviewed during periodic evaluations of the safety management system?**

A. Effectiveness of controls, audit findings, and achievement of safety objectives
B. Payroll and HR files
C. Employee leisure schedules
D. Customer satisfaction reports

**Q304. Why is it important to include safety criteria in procurement decisions?**

A. It reduces competition unfairly
B. It adds unnecessary cost to bidding
C. It ensures purchased materials, tools, and services meet safety and compliance requirements
D. It delays project delivery

**Q305. What factor most influences the long-term effectiveness of a construction safety program?**

A. Management commitment combined with employee engagement and continuous improvement
B. Short-term compliance campaigns
C. Enforcement through penalties only
D. Restricting safety to top management levels

**Q306. What is the most important factor in determining the effectiveness of a site safety committee?**

A. Frequency of meetings
B. Active participation and follow-through on identified actions
C. Number of management representatives
D. Length of agendas

**Q307. Which step should immediately follow the identification of a serious safety hazard during an audit?**

A. Documenting and reporting the issue for corrective action
B. Waiting until the audit is completed
C. Ignoring the issue if no incident occurs
D. Informing only upper management after project completion

**Q308. Why is a written safety policy essential for a construction organization?**

A. It replaces the need for jobsite supervision
B. It formalizes leadership commitment and defines safety expectations
C. It satisfies marketing purposes only
D. It limits flexibility in decision-making

**Q309. Which approach ensures the accuracy of incident data used for trend analysis?**

A. Consistent reporting procedures and verification of data sources
B. Recording only lost-time injuries
C. Allowing supervisors to submit reports selectively
D. Using estimates instead of actual data

**Q310. What is the main purpose of conducting periodic safety management reviews?**
A. To reduce audit frequency
B. To prepare for regulatory inspections only
C. To replace daily supervision
D. To evaluate performance, identify gaps, and promote continual improvement
**Q311. How can safety performance metrics support strategic decision-making?**
A. By providing data-driven insights into program effectiveness and resource needs
B. By replacing leadership judgment entirely
C. By focusing only on financial outcomes
D. By discouraging further investment in safety
**Q312. What distinguishes an effective corrective action process after an audit?**
A. Documented actions, assigned responsibilities, and verification of completion
B. Generic recommendations without accountability
C. Deferring actions indefinitely
D. Ignoring minor findings to save time
**Q313. Which control strategy should be prioritized when developing safety programs?**
A. Relying primarily on PPE use
B. Eliminating hazards through design or substitution
C. Scheduling frequent toolbox talks
D. Documenting hazards without correction
**Q314. How does management demonstrate visible commitment to safety initiatives?**
A. Participating in field inspections, providing resources, and recognizing safe performance
B. Delegating all responsibilities to safety staff
C. Issuing memos without engagement
D. Focusing solely on production goals
**Q315. What is a key characteristic of a mature safety management system?**
A. Annual policy reviews without action
B. Reactive response to incidents only
C. Integration of safety into all business processes with continuous feedback loops
D. Separation of safety from operational planning
**Q316. Why is contractor onboarding critical in multi-employer worksites?**
A. It establishes safety expectations and ensures alignment with site requirements before work begins
B. It delays work unnecessarily
C. It serves only as a compliance formality
D. It transfers all responsibility to the general contractor
**Q317. Which type of safety performance indicator measures proactive effort rather than outcomes?**
A. Leading indicators such as inspections and near-miss reports
B. Lagging indicators like injury rates
C. Workers' compensation costs
D. Lost-time frequency rates
**Q318. What is the main purpose of a safety performance dashboard?**
A. To eliminate the need for reports
B. To store unused data
C. To replace direct communication with employees
D. To provide real-time visualization of key metrics for informed decision-making
**Q319. Why is workforce involvement crucial to sustaining safety program success?**
A. It ensures ownership, practical feedback, and engagement in continuous improvement
B. It increases management workload only

C. It delays decision-making
D. It diminishes accountability

**Q320. What is the final objective of implementing and sustaining a safety management system?**
A. To achieve continuous improvement and a proactive culture preventing workplace incidents
B. To comply only with minimum legal requirements
C. To maintain documentation for audits only
D. To assign responsibility solely to safety officers

**Q321. What is the most important characteristic of an effective safety leader?**
A. Technical knowledge
B. Authority and control
C. Integrity and consistency
D. Strict enforcement of rules

**Q322. Which method best encourages worker participation in safety programs?**
A. Mandating attendance at safety meetings
B. Rewarding proactive safety suggestions
C. Assigning additional duties for compliance
D. Reducing feedback opportunities

**Q323. A foreman notices that crew members ignore new safety signage. Which leadership approach should he take first?**
A. Enforce disciplinary action immediately
B. Report the incident to senior management
C. Increase signage throughout the site
D. Reassess the clarity and visibility of the signage

**Q324. Which communication practice most strengthens safety culture?**
A. Top-down announcements only
B. Two-way communication between workers and supervisors
C. Weekly reports emailed to staff
D. Posting policy memos on notice boards

**Q325. Which leadership style most effectively promotes long-term safety performance?**
A. Authoritarian leadership
B. Transactional leadership
C. Transformational leadership
D. Laissez-faire leadership

**Q326. What is the most effective way to motivate employees to follow safety protocols?**
A. Threatening disciplinary action
B. Explaining the personal benefits of safety compliance
C. Offering cash rewards only
D. Posting safety statistics in common areas

**Q327. A safety manager conducts a meeting where workers identify solutions to recurring hazards. This represents which leadership trait?**
A. Delegation
B. Empowerment
C. Command
D. Compliance

**Q328. During a safety meeting, a supervisor dominates discussion and discourages worker input. What leadership weakness does this show?**
A. Lack of communication clarity
B. Poor listening skills

C. Weak technical competence
D. Excessive delegation

**Q329. Which training method is most effective for teaching high-risk construction tasks?**
A. Lecture-based classroom instruction
B. Self-paced online courses
C. Hands-on practical demonstrations
D. Reading safety manuals

**Q330. What is the first step in developing a safety training program?**
A. Writing the lesson plan
B. Identifying training needs
C. Conducting a post-training evaluation
D. Purchasing training materials

**Q331. Which approach helps ensure trainees retain and apply safety information?**
A. Frequent testing
B. Extended lectures
C. Interactive, participatory training sessions
D. Handouts with written policies

**Q332. Which behavior best demonstrates effective communication by a safety leader?**
A. Speaking frequently about safety policies
B. Providing clear, concise, and actionable feedback
C. Writing detailed safety manuals
D. Reporting all hazards to management

**Q333. A supervisor consistently praises safe behavior in team meetings. What effect does this have on safety culture?**
A. It reinforces positive behaviors and engagement
B. It discourages open communication
C. It reduces reporting of incidents
D. It creates dependency on rewards

**Q334. Which factor most influences the success of safety leadership training?**
A. Trainer's technical expertise
B. Trainer's ability to connect and engage with learners
C. Duration of the training session
D. Quantity of materials provided

**Q335. Which component of a safety training program measures behavioral improvement after completion?**
A. Pre-training survey
B. Observation and follow-up evaluation
C. Participant satisfaction questionnaire
D. Attendance log

**Q336. A manager regularly meets with employees to discuss upcoming tasks and potential hazards. This practice primarily supports which leadership principle?**
A. Command authority
B. Technical supervision
C. Disciplinary control
D. Collaborative communication

**Q337. Which behavior by management most undermines safety culture?**
A. Enforcing PPE use consistently
B. Ignoring minor safety violations by senior staff

C. Conducting regular toolbox talks
D. Providing timely feedback

**Q338. What should a safety leader do when workers resist new safety procedures?**

A. Enforce penalties immediately
B. Explain the reasons and benefits behind the change
C. Delay implementation until agreement
D. Ignore resistance to avoid conflict

**Q339. What is the main purpose of a safety observation program?**

A. Record compliance statistics
B. Evaluate management performance
C. Identify unsafe acts and reinforce safe behaviors
D. Replace formal training sessions

**Q340. Which leadership activity best promotes continuous safety improvement?**

A. Conducting investigations only after incidents
B. Encouraging employee suggestions and acting on them
C. Reducing meeting frequency
D. Focusing only on compliance audits

**Q341. What is the key objective of behavior-based safety (BBS) programs?**

A. Eliminating all workplace hazards
B. Identifying and reinforcing safe behaviors
C. Replacing engineering controls
D. Reducing paperwork requirements

**Q342. Which communication tool is most effective for delivering critical safety updates on short notice?**

A. Monthly newsletter
B. Email campaign
C. Annual safety conference
D. Toolbox talks

**Q343. In safety leadership, what does the term "visible commitment" mean?**

A. Attending safety meetings occasionally
B. Demonstrating safety behaviors consistently on-site
C. Reviewing reports from the office
D. Delegating safety to supervisors

**Q344. A company launches a mentorship program where experienced workers coach new hires on safety practices. What leadership principle does this reflect?**

A. Discipline enforcement
B. Knowledge transfer and capacity building
C. Authority delegation
D. Production supervision

**Q345. Which behavior by supervisors best promotes psychological safety within teams?**

A. Allowing workers to report mistakes without fear
B. Strictly penalizing all violations
C. Focusing communication only on results
D. Avoiding feedback to prevent conflict

**Q346. Which communication skill is most important when delivering safety training to a multilingual workforce?**

A. Speaking louder for clarity
B. Using visual aids and demonstrations
C. Providing printed materials only

D. Reducing interaction to avoid confusion

**Q347. What is the main goal of safety communication during toolbox talks?**

A. Reporting project progress
B. Sharing operational targets
C. Reinforcing safety awareness and procedures
D. Discussing employee attendance

**Q348. When an employee reports a near miss, what is the best leadership response?**

A. Ignore it if no one was injured
B. Thank the employee and review preventive actions
C. File the report without acknowledgment
D. Punish the employee for carelessness

**Q349. Which of the following best illustrates proactive safety leadership?**

A. Waiting for incidents to review procedures
B. Regularly walking worksites and engaging workers
C. Delegating inspections to lower staff
D. Relying solely on monthly reports

**Q350. What is a common barrier to effective safety communication?**

A. Clear instruction delivery
B. Cultural and language differences
C. Visual reinforcement
D. Feedback collection

**Q351. What should a leader do when introducing a new safety initiative?**

A. Announce it without explanation
B. Implement it quietly to avoid questions
C. Mandate compliance immediately
D. Involve employees early and explain objectives

**Q352. Which factor most influences adult learning in safety training?**

A. Length of presentation
B. Relevance to job experience
C. Technical jargon
D. Volume of materials

**Q353. What leadership behavior most enhances a team's trust in safety management?**

A. Assigning all decisions to upper management
B. Demonstrating transparency and accountability
C. Avoiding communication about safety issues
D. Prioritizing production over protection

**Q354. Which statement describes the role of emotional intelligence in safety leadership?**

A. It reduces the need for communication
B. It helps leaders understand and manage workforce emotions
C. It encourages strict enforcement of rules only
D. It replaces technical safety knowledge

**Q355. What is the purpose of a communication audit in a safety program?**

A. To evaluate effectiveness of safety messages and channels
B. To identify top-performing employees
C. To analyze production targets
D. To assess financial metrics

**Q356. Which practice best supports long-term retention of safety knowledge?**

A. One-time seminars

B. Printed rulebooks
C. Refresher training and follow-up assessments
D. Relying on experience alone

**Q357. What leadership action most effectively develops a positive safety culture?**
A. Enforcing discipline strictly
B. Engaging employees in safety decision-making
C. Prioritizing productivity goals
D. Delegating all safety duties

**Q358. How can leaders effectively evaluate the impact of safety communication campaigns?**
A. By measuring worker understanding and behavioral change
B. By counting the number of emails sent
C. By checking attendance at meetings
D. By reviewing financial budgets

**Q359. What is the first step in developing a leadership-driven safety culture?**
A. Setting clear safety values and vision
B. Increasing production rates
C. Implementing audits immediately
D. Focusing on short-term incentives

**Q360. How should a supervisor address an employee's repeated minor safety violations?**
A. Issue immediate suspension
B. Discuss the reasons and coach corrective behavior
C. Ignore to maintain goodwill
D. Report directly to upper management

**Q361. What is the most effective way to communicate emergency procedures to new employees?**
A. Through written manuals only
B. Via orientation sessions with demonstrations
C. By posting evacuation maps
D. Through annual review meetings

**Q362. What characteristic defines a learning organization in safety management?**
A. Continuous improvement through shared knowledge
B. Rigid adherence to outdated procedures
C. Limiting feedback to management only
D. Avoiding discussion of incidents

**Q363. Which leadership approach is most effective for improving team communication?**
A. Directive
B. Isolated
C. Passive
D. Collaborative

**Q364. Which factor best predicts successful safety leadership implementation?**
A. Consistent leadership behavior aligning with stated values
B. Length of safety meetings
C. Number of safety posters displayed
D. Quantity of policies written

**Q365. What should be the focus of feedback after observing unsafe behavior?**
A. Identifying blame
B. Providing constructive coaching and solutions
C. Discussing unrelated issues
D. Reporting the employee to HR

**Q366. What distinguishes a proactive safety culture from a reactive one?**
A. Taking preventive measures before incidents occur
B. Responding only after accidents happen
C. Reducing inspection frequency
D. Waiting for audits to reveal issues
**Q367. Which leadership behavior best supports team accountability?**
A. Setting clear expectations and following up consistently
B. Allowing informal rule flexibility
C. Avoiding difficult conversations
D. Delegating responsibility without oversight
**Q368. What is the role of recognition programs in developing safety culture?**
A. To promote competition among workers
B. To reinforce desired safety behaviors positively
C. To replace disciplinary systems
D. To emphasize productivity over safety
**Q369. When communicating complex safety topics, which strategy ensures understanding?**
A. Using simple language and examples
B. Including as much technical jargon as possible
C. Reading from manuals directly
D. Limiting audience participation
**Q370. What is the primary goal of a safety leadership mentorship program?**
A. To enforce discipline
B. To replace management responsibilities
C. To develop leadership competence in safety practices
D. To reduce training costs
**Q371. What is the most effective way to evaluate the success of safety leadership initiatives?**
A. Reviewing policy documents
B. Measuring changes in safety behavior and performance
C. Counting the number of meetings held
D. Tracking completed inspections only
**Q372. How should a leader handle conflicting priorities between production and safety?**
A. Prioritize safety and communicate its long-term value
B. Allow production to continue if risks are moderate
C. Defer safety decisions until after project completion
D. Ignore minor safety breaches for efficiency
**Q373. What is the role of leadership visibility during safety inspections?**
A. Symbolic presence only
B. Encouraging open dialogue and engagement with workers
C. Supervising technical details exclusively
D. Completing paperwork for compliance
**Q374. Which of the following best demonstrates servant leadership in safety management?**
A. Making decisions without employee input
B. Prioritizing worker needs and removing barriers to safe performance
C. Delegating all safety tasks to supervisors
D. Enforcing rules without flexibility
**Q375. What is a key indicator of strong communication culture within an organization?**
A. Employees freely report hazards and near misses
B. Reports only come from supervisors

C. Safety feedback flows in one direction
D. Incidents are rarely discussed

**Q376. What is the most effective way to ensure learning after an incident investigation?**
A. Archive the report for future reference
B. Avoid public discussion to protect reputations
C. Discuss only with senior management
D. Share findings openly and implement corrective training

**Q377. What is the most likely consequence of inconsistent leadership communication?**
A. Strengthened employee trust
B. Confusion and reduced compliance
C. Increased teamwork
D. Better understanding of safety goals

**Q378. How can storytelling enhance safety communication?**
A. By presenting rules as narratives with emotional relevance
B. By adding humor to distract from serious content
C. By focusing only on management success stories
D. By using overly technical descriptions

**Q379. What is the purpose of peer-to-peer safety observation programs?**
A. To assign blame for unsafe acts
B. To foster teamwork and continuous improvement
C. To reduce management workload
D. To replace professional safety inspections

**Q380. What type of feedback most effectively corrects unsafe behavior?**
A. Immediate, specific, and supportive feedback
B. Delayed, general feedback
C. Public criticism
D. Written warnings only

**Q381. Which leadership action best builds a safety-focused identity among workers?**
A. Minimizing talk about safety during meetings
B. Emphasizing production achievements
C. Encouraging pride in working safely
D. Outsourcing safety leadership

**Q382. Which method best ensures that safety messages are understood across all organizational levels?**
A. Tailoring communication to the audience's literacy and role
B. Using identical technical content for everyone
C. Communicating only through digital channels
D. Limiting repetition of key messages

**Q383. What is the role of empathy in safety leadership?**
A. Reducing discipline for safety violations
B. Understanding worker challenges to provide meaningful support
C. Replacing rules with emotional appeals
D. Avoiding accountability conversations

**Q384. Why is active participation in safety meetings important for supervisors?**
A. It demonstrates leadership engagement and sets behavioral standards
B. It helps complete administrative requirements
C. It allows delegation of responsibilities
D. It minimizes field work expectations

**Q385. What is the key purpose of post-training evaluations?**

A. Meeting legal audit requirements
B. Recording attendance for compliance
C. Reducing the need for refresher courses
D. Measuring knowledge transfer and identifying improvement areas

**Q386. What behavior demonstrates "leading by example" in a construction setting?**
A. Always wearing required PPE correctly on-site
B. Issuing safety directives from the office
C. Allowing exceptions for senior workers
D. Discussing safety once per month

**Q387. Which approach best addresses communication breakdowns between teams?**
A. Conducting structured feedback sessions to clarify misunderstandings
B. Assigning blame to the most vocal team
C. Restricting communication between departments
D. Ignoring small issues until escalation

**Q388. What is the most effective way to promote diversity and inclusion in safety communication?**
A. Using identical training for every department
B. Assuming all workers understand English
C. Translating materials and using culturally sensitive examples
D. Avoiding discussions about cultural differences

**Q389. What does the term "safety climate" refer to?**
A. Workers' shared perceptions of management's commitment to safety
B. The frequency of safety meetings
C. The number of incidents in the past year
D. Environmental conditions on the job site

**Q390. Which leadership communication method best prevents rumor and misinformation after an incident?**
A. Immediate, transparent updates from credible sources
B. Waiting for investigations to finish completely
C. Relying on unofficial channels for news
D. Minimizing information sharing

**Q391. What is the main benefit of cross-functional safety committees?**
A. Reducing management workload
B. Encouraging collaboration between departments and improving hazard awareness
C. Increasing documentation requirements
D. Limiting employee input

**Q392. How can leaders ensure training effectiveness over time?**
A. Scheduling periodic refresher sessions and real-world assessments
B. Conducting one initial training only
C. Avoiding evaluation to save time
D. Posting online reminders occasionally

**Q393. Which communication behavior most supports open reporting of near misses?**
A. Responding positively and without blame
B. Criticizing the worker for carelessness
C. Ignoring minor reports
D. Focusing solely on injury cases

**Q394. Which of the following is an example of transformational leadership in safety?**
A. Prioritizing paperwork over people
B. Enforcing compliance through punishment

C. Minimizing engagement to reduce workload
D. Inspiring workers to see safety as a shared vision and value

**Q395. How does coaching differ from supervision in safety leadership?**
A. Coaching focuses on development and guidance, not enforcement
B. Coaching replaces all management oversight
C. Supervision eliminates the need for training
D. Both are identical in purpose

**Q396. What is the most effective way for leaders to maintain momentum in long-term safety initiatives?**
A. Regularly reviewing goals and celebrating milestones
B. Issuing a single launch announcement
C. Relying solely on written progress reports
D. Shifting focus to new priorities frequently

**Q397. How can safety leaders strengthen workers' sense of ownership for site safety?**
A. Centralizing all decision-making in management
B. Assigning them observation duties without consultation
C. Involving them in hazard assessments and solution development
D. Limiting their input to reduce delays

**Q398. What leadership behavior most effectively supports continuous improvement after audits?**
A. Acting promptly on findings and involving teams in corrective planning
B. Filing audit results for recordkeeping only
C. Delaying actions until the next inspection
D. Limiting access to audit outcomes

**Q399. In promoting safety leadership, why is credibility critical?**
A. Employees trust and follow leaders whose actions match their words
B. It allows leaders to avoid accountability for mistakes
C. It reduces the need for consistent communication
D. It emphasizes authority over example

**Q400. What is the ultimate goal of leadership, communication, and training in safety management?**
A. Achieving compliance through enforcement
B. Building a sustainable culture of safety ownership and excellence
C. Meeting legal and audit requirements
D. Reducing training costs

---

## Answer keys and Explanations (301-400)

**Q301. Answer: D**

**Explanation**: Structured meetings analyze metrics, discuss recent incidents, and develop targeted corrective actions. They maintain transparency and decision-making continuity. Casual meetings lack direction and accountability. Systematic discussions foster collective problem-solving, clear follow-up, and leadership engagement—hallmarks of mature safety management systems driving measurable performance improvement and teamwork.

**Q302. Answer: A**

**Explanation**: Toolbox talks reinforce expectations, addressing current hazards and fostering open communication. Frequent, interactive briefings encourage participation and ensure continuous awareness. Static postings or sporadic meetings fail to maintain engagement. Regular, two-way discussions ensure understanding, commitment, and consistent application of company safety standards across all crews.

**Q303. Answer: A**

**Explanation**: Evaluating control effectiveness and reviewing performance against established goals ensures the system functions as intended. Integrating audit data reveals trends requiring attention. Administrative reviews provide no improvement insight. Structured evaluations drive corrective action and continuous development of an effective, adaptive safety management framework organization-wide.

**Q304. Answer: C**

**Explanation**: Safety-conscious procurement integrates hazard prevention at the acquisition stage, ensuring materials and contractors conform to established standards. This reduces downstream risk from unsafe equipment or suppliers. Incorporating safety in procurement processes demonstrates organizational responsibility, regulatory alignment, and commitment to lifecycle safety and operational integrity.

**Q305. Answer: A**

**Explanation**: Sustainable safety excellence arises when leadership visibly supports safety and workers actively participate. Combined engagement ensures shared accountability, constant learning, and innovation. Short-term enforcement alone cannot sustain progress. Collaborative commitment and adaptive improvement processes embed safety into organizational DNA, creating resilience and consistent performance in dynamic construction environments.

**Q306. Answer: B**

**Explanation**: The true measure of a safety committee's effectiveness lies in its ability to engage members, identify hazards, and ensure corrective actions are completed. Frequent meetings alone do not guarantee progress. Active participation fosters ownership, while consistent follow-through demonstrates accountability and commitment to improvement. Effective committees drive communication, enhance compliance, and continuously strengthen the site's overall safety culture through shared responsibility and visible results across all project levels.

**Q307. Answer: A**

**Explanation**: Immediate documentation and communication of serious hazards prevent potential injuries by ensuring rapid intervention. Waiting for report completion delays response and increases risk. Proper follow-up requires assigning responsibility and verifying corrective measures. Addressing issues promptly demonstrates diligence, compliance, and proactive hazard management, reinforcing a culture where safety priorities outweigh administrative convenience and protecting both workers and project integrity effectively.

**Q308. Answer: B**

**Explanation**: A written safety policy communicates the organization's values, responsibilities, and objectives related to health and safety. It provides structure, ensuring consistent expectations across projects. Without a formal policy, enforcement and accountability weaken. This foundational document supports compliance, demonstrates management's commitment, and guides behavior at every level, anchoring the company's culture of prevention and continuous improvement in construction environments.

**Q309. Answer: A**

**Explanation**: Reliable data stems from standardized reporting and verification, ensuring completeness and accuracy. Selective or inconsistent reporting distorts analysis, leading to poor decision-making. Accurate data enables identification of genuine trends, supports targeted interventions, and validates program effectiveness. Verifying information through audits and cross-referencing fosters credibility and ensures management bases safety strategies on factual, comprehensive evidence rather than assumptions or incomplete inputs.

**Q310. Answer: D**

**Explanation**: Periodic management reviews ensure alignment between goals and outcomes by analyzing performance data, audit results, and compliance status. This structured evaluation identifies opportunities for process enhancement and resource allocation. Reviews are proactive tools for continuous improvement, not mere compliance checks. They sustain organizational learning, ensuring that safety systems evolve alongside operational and regulatory developments to maintain consistent, high-level safety performance.

**Q311. Answer: A**

**Explanation**: Safety metrics translate qualitative activities into measurable results, allowing management to assess effectiveness, prioritize investments, and justify resource allocation. Data-driven analysis identifies strengths and weaknesses, supporting informed decision-making. Reducing safety to financial terms overlooks its strategic value. Metrics bridge operational reality with leadership vision, enabling continuous refinement and alignment of safety goals with organizational performance objectives and sustainability priorities.

**Q312. Answer: A**

**Explanation**: A structured corrective action process ensures that identified deficiencies are addressed systematically. Assigning ownership and verifying implementation transforms findings into tangible improvements. Generic or deferred actions reduce credibility and allow recurring issues. Verification through follow-up audits confirms effectiveness, reinforcing accountability and fostering an organizational culture committed to continuous, measurable enhancement of safety practices and compliance performance.

**Q313. Answer: B**

**Explanation**: Hazard elimination through design or substitution represents the most effective and permanent control method in the hierarchy of controls. Engineering out hazards reduces reliance on behavior-based measures. PPE is essential but last in priority. Proactive elimination prevents exposure, reduces risk at the source, and simplifies long-term management of safety, aligning with best practice and regulatory expectations for prevention-first strategies.

**Q314. Answer: A**

**Explanation**: Direct management involvement sends a powerful message that safety is a top priority. Field participation bridges leadership with workforce realities, while adequate resources and recognition sustain motivation. Passive or symbolic gestures fail to inspire change. Authentic engagement fosters trust, accountability, and shared ownership of safety outcomes across the organization, embedding prevention in daily operations and leadership behavior.

**Q315. Answer: C**

**Explanation**: Mature systems integrate safety into planning, procurement, and performance evaluation, supported by ongoing data analysis and stakeholder engagement. This approach ensures risks are identified early and addressed holistically. Reactive or isolated systems fail to evolve. Continuous feedback ensures adaptability and sustainability, reinforcing safety as an operational core value and strategic advantage for long-term business success.

**Q316. Answer: A**

**Explanation**: Contractor onboarding introduces site-specific rules, hazards, and procedures, ensuring all parties understand their responsibilities before mobilization. This alignment prevents miscommunication and inconsistent practices. Treating onboarding as a formality undermines safety coordination. A robust process builds collaboration and accountability, strengthening control over shared work environments and maintaining unified compliance and performance across project partners.

**Q317. Answer: A**

**Explanation**: Leading indicators capture proactive activities that prevent incidents—like audits, observations, and safety training participation. These metrics predict potential risks and guide timely interventions. Lagging indicators measure past harm. Prioritizing leading metrics shifts focus to prevention, improving engagement and overall program effectiveness while building resilience and proactive safety culture maturity across operations.

**Q318. Answer: D**

**Explanation**: A dashboard consolidates performance data into visual indicators, enabling timely, data-driven decisions. It identifies trends, tracks progress toward objectives, and highlights emerging risks. Static data lacks immediacy and context. Dashboards enhance transparency, accountability, and management responsiveness, empowering organizations to adjust strategies quickly and sustain performance excellence in safety and risk management.

**Q319. Answer: A**

**Explanation**: Worker participation brings real-world insights into identifying hazards and verifying control effectiveness. It promotes shared responsibility, trust, and commitment to program goals. Without engagement, safety remains a top-down directive prone to resistance. Collaborative participation transforms compliance into a cultural norm, fostering continuous improvement, empowerment, and organizational unity in achieving zero-harm objectives consistently and effectively.

**Q320. Answer: A**

**Explanation**: The ultimate goal of a safety management system is to create a proactive, learning-oriented organization where continuous improvement eliminates hazards and prevents incidents. Compliance serves as a baseline, not the endpoint. Sustained systems embed safety into every decision, promoting innovation, efficiency, and wellbeing while enhancing productivity and organizational reputation through consistent excellence and leadership commitment.

**Q321. Answer: C**

**Explanation**: Effective safety leaders exhibit integrity and consistency in their actions, fostering trust and credibility among workers. When leaders model the behavior they expect, employees are more likely to follow safety policies voluntarily. Technical expertise and authority are helpful, but without moral consistency, leadership influence is limited in sustaining long-term safety culture improvements.

**Q322. Answer: B**

**Explanation**: Recognizing and rewarding proactive safety suggestions motivates workers to engage in continuous improvement. This approach enhances ownership and accountability within the safety program. Mandates alone may generate compliance without commitment, while removing feedback channels undermines open communication vital for engagement.

**Q323. Answer: D**

**Explanation**: Reviewing the clarity, placement, and language of safety signage ensures workers understand and notice critical information. Often, communication barriers—such as poor visibility or unclear wording—cause noncompliance. Effective leaders evaluate communication quality before enforcing discipline, focusing on understanding and improvement first.

**Q324. Answer: B**

**Explanation**: Two-way communication allows employees to voice concerns, report hazards, and share improvement ideas, creating mutual respect and ownership. It transforms safety from a directive to a dialogue. Static memos or one-way reports lack the engagement and feedback necessary for active safety culture development.

**Q325. Answer: C**

**Explanation**: Transformational leaders inspire and empower their teams to value safety intrinsically rather than through external enforcement. They emphasize vision, trust, and motivation. Authoritarian or transactional styles can ensure compliance short-term but rarely instill genuine commitment to safe behaviors across the organization.

**Q326. Answer: B**

**Explanation**: Communicating the personal and team benefits of safety connects motivation to intrinsic values such as well-being and teamwork. It builds commitment based on understanding rather than fear. Monetary rewards may create short-lived motivation, while punitive tactics reduce morale and participation.

**Q327. Answer: B**

**Explanation**: Empowerment involves engaging employees in decision-making, encouraging ownership and accountability for safety improvements. Workers who contribute to problem-solving are more likely to adopt safe behaviors. It transforms leadership from top-down control into collaborative influence, a core feature of positive safety culture.

**Q328. Answer: B**

**Explanation**: Effective safety leaders practice active listening to understand worker concerns and ideas. Ignoring or dominating discussions creates disengagement and mistrust. Open dialogue fosters collaboration and continuous improvement, making listening as important as speaking in leadership communication.

**Q329. Answer: C**

**Explanation**: Practical, hands-on training provides experiential learning that reinforces proper techniques and hazard recognition. It ensures skills are retained and applied in real situations. While lectures and online courses provide foundational knowledge, physical practice is essential for high-risk operational proficiency.

**Q330. Answer: B**

**Explanation**: Identifying training needs ensures content is relevant to actual workplace hazards and skill gaps. This assessment guides all subsequent design, delivery, and evaluation stages. Skipping this step risks creating generic or ineffective programs that fail to meet organizational safety requirements.

**Q331. Answer: C**

**Explanation**: Interactive sessions engage learners through discussion, simulation, and practice, improving retention and comprehension. Adult learners benefit from participation that connects theory with real-life application. Passive methods like long lectures or handouts do not reinforce behavioral learning effectively.

**Q332. Answer: B**

**Explanation**: Clear and actionable feedback helps workers understand expectations, correct errors, and reinforce good practices. Communication must be concise, timely, and supportive. Merely talking about safety or producing documents does not ensure understanding or behavioral improvement.

**Q333. Answer: A**

**Explanation**: Recognition of safe behavior reinforces the value of safety and encourages repetition of positive actions. Positive reinforcement enhances engagement and morale, making employees more likely to follow safety practices consistently. It supports proactive rather than reactive safety management.

**Q334. Answer: B**

**Explanation**: Trainers who connect with and engage learners foster participation and retention. Engagement ensures learners internalize content and apply it on-site. Technical expertise is valuable but ineffective if not communicated clearly and relatable to trainees' experiences.

**Q335. Answer: B**

**Explanation**: Post-training observation and evaluation assess how well trainees apply knowledge in their work environment. This behavioral measurement reflects the program's real-world effectiveness. Attendance or satisfaction data only show participation, not actual performance change.

**Q336. Answer: D**

**Explanation**: Collaborative communication fosters trust, participation, and joint problem-solving between leaders and employees. It demonstrates that safety leadership is about partnership and prevention rather than control. Regular dialogue creates openness and shared accountability for hazard management.

**Q337. Answer: B**

**Explanation**: Inconsistent enforcement of safety rules erodes credibility and trust, signaling double standards. Employees quickly notice unfair practices, which reduce engagement and compliance. Consistency is essential for a culture where safety rules apply equally to all members of the organization.

**Q338. Answer: B**

**Explanation**: Explaining the rationale and benefits helps employees understand the purpose of the new procedure, reducing resistance. Effective leaders use communication and involvement to foster acceptance. Penalties or avoidance increase resistance, while delay undermines leadership credibility.

**Q339. Answer: C**

**Explanation**: Observation programs allow supervisors to identify unsafe behaviors early and coach employees toward safer actions. They create opportunities for immediate feedback and positive reinforcement. Their purpose is not data collection alone but active engagement in continuous improvement.

**Q340. Answer: B**

**Explanation**: Continuous improvement thrives on employee input and management responsiveness. Encouraging suggestions creates shared ownership of safety performance. Restricting focus to audits or reactive measures limits innovation and engagement, hindering long-term development of safety excellence.

**Q341. Answer: B**

**Explanation**: BBS programs focus on observing, analyzing, and reinforcing safe behaviors to reduce human error and promote consistent safety practices. They complement hazard control methods by addressing behavioral causes of incidents. The goal is improved performance through awareness and feedback.

**Q342. Answer: D**

**Explanation**: Toolbox talks provide immediate, focused communication that reaches field workers directly before tasks begin. They enable timely delivery of updates or procedural changes. Newsletters and conferences are useful for awareness but lack immediacy and practical application.

**Q343. Answer: B**

**Explanation**: Visible commitment means leaders actively model and demonstrate safe behavior in daily operations. It builds credibility and signals that safety is a personal priority. Passive oversight or delegation alone fails to inspire employees to adopt safety-minded behaviors.

**Q344. Answer: B**

**Explanation**: Mentorship enhances learning through experience sharing, improving competence and confidence in new workers. It reinforces leadership through guidance and example, ensuring continuity in safety standards. This peer-based approach strengthens organizational learning and culture.

**Q345. Answer: A**

**Explanation**: Encouraging open reporting without fear supports psychological safety, where workers feel comfortable raising concerns and learning from mistakes. This openness improves trust, teamwork, and proactive problem-solving. Excessive punishment or avoidance suppresses communication and increases hidden risks.

**Q346. Answer: B**

**Explanation**: Visual aids and demonstrations transcend language barriers, ensuring comprehension across diverse groups. They reinforce key messages through visual learning, which is universal and effective for practical topics. Speaking louder or minimizing interaction doesn't solve comprehension issues and can alienate participants.

**Q347. Answer: C**

**Explanation**: Toolbox talks reinforce safety awareness, highlight current hazards, and refresh procedural understanding before work begins. They are essential for short, focused, and practical communication. Project progress or attendance discussions are secondary and not the main goal of these sessions.

**Q348. Answer: B**

**Explanation**: Acknowledging and thanking employees encourages continued reporting and proactive risk management. Reviewing corrective actions ensures lessons are learned and hazards mitigated. Ignoring or punishing discourages transparency and undermines trust within the safety culture.

**Q349. Answer: B**

**Explanation**: Regular engagement with workers through site walks shows proactive commitment and direct involvement in identifying risks. It builds rapport, trust, and early intervention. Delegating or depending on reports alone creates distance between leadership and the actual safety environment.

**Q350. Answer: B**

**Explanation**: Cultural and language differences can hinder mutual understanding, leading to misinterpretation of safety instructions. Overcoming this requires adapting messages through translation, visuals, and inclusivity. Clear instructions and visual reinforcement enhance communication, not impede it.

**Q351. Answer: D**

**Explanation**: Involving employees early fosters buy-in, ownership, and smoother implementation. When workers understand the reasoning and goals, they are more motivated to participate. Top-down mandates often create resistance, while exclusion undermines engagement and trust.

**Q352. Answer: B**

**Explanation**: Adults learn best when material directly relates to their work experience and real-life applications. Practical relevance improves engagement and retention. Overly long or technical sessions decrease focus and hinder meaningful learning outcomes.

**Q353. Answer: B**

**Explanation**: Transparency and accountability build trust by showing consistency and honesty in safety-related actions. When leaders take responsibility and communicate openly, teams feel respected and supported. Concealment or conflicting priorities erodes confidence in leadership.

**Q354. Answer: B**

**Explanation**: Emotional intelligence allows leaders to empathize with workers, manage conflict, and respond constructively to stress. Recognizing emotions strengthens team morale and cooperation. It complements technical knowledge and enhances interpersonal effectiveness, critical in safety-sensitive environments.

**Q355. Answer: A**

**Explanation**: Communication audits assess how well safety messages reach, engage, and influence the workforce. They help identify barriers, measure understanding, and guide improvements. Production or financial data are unrelated to the communication flow and safety perception analysis.

**Q356. Answer: C**

**Explanation**: Refresher training reinforces learning, prevents skill decay, and ensures knowledge remains current. Periodic review and assessment keep safety principles active in employees' minds. One-time sessions or passive materials are insufficient for sustained behavioral change.

**Q357. Answer: B**

**Explanation**: Inclusion in safety decisions enhances ownership, accountability, and morale. When workers contribute to identifying hazards and solutions, they align more closely with organizational safety goals. Strict discipline without involvement often leads to fear and minimal compliance.

**Q358. Answer: A**

**Explanation**: Evaluating communication effectiveness involves assessing comprehension, recall, and observed behavioral changes. Impact is shown when messages lead to safer practices. Quantity-based metrics like email counts or attendance do not reflect actual effectiveness.

**Q359. Answer: A**

**Explanation**: A strong safety culture begins with clearly defined values and a shared vision. This foundation guides decision-making and behavior across the organization. Without established principles, subsequent actions lack direction and sustainability.

**Q360. Answer: B**

**Explanation**: Coaching provides understanding and guidance, addressing root causes like misunderstanding or complacency. It maintains respect and encourages improvement. Immediate punishment or avoidance prevents behavioral correction and weakens team cohesion.

**Q361. Answer: B**

**Explanation**: Orientation sessions with demonstrations ensure hands-on learning and immediate understanding of emergency actions. Visual and interactive methods strengthen retention and confidence. Written materials or signage serve as references but lack the impact of active learning.

**Q362. Answer: A**

**Explanation**: Learning organizations use feedback, analysis, and collaboration to adapt and improve continuously. They view mistakes as learning opportunities, not failures. This adaptability fosters innovation and sustained safety excellence.

**Q363. Answer: D**

**Explanation**: Collaborative leadership invites participation, dialogue, and shared responsibility. It enhances team cohesion and clarity, ensuring communication is open and bidirectional. Directive or passive styles reduce engagement and slow feedback flow.

**Q364. Answer: A**

**Explanation**: Alignment between words and actions builds credibility and reinforces trust in leadership. When leaders practice what they promote, employees model similar behavior. Symbolic actions like posters or lengthy meetings have little influence without behavioral consistency.

**Q365. Answer: B**

**Explanation**: Constructive coaching focuses on solutions, encouraging employees to understand and correct behavior without fear. This fosters improvement and commitment to safety. Blame or unrelated discussion diminishes learning and trust.

**Q366. Answer: A**

**Explanation**: Proactive cultures identify and manage risks early through observation, analysis, and open communication. They prioritize prevention over response, improving overall performance. Reactive cultures rely on corrective actions after events, missing early intervention opportunities.

**Q367. Answer: A**

**Explanation**: Clear expectations define roles, while consistent follow-up ensures accountability and performance. It demonstrates fairness and reliability. Avoidance or lack of clarity weakens standards and can erode discipline.

**Q368. Answer: B**

**Explanation**: Recognition programs encourage repetition of positive behaviors by acknowledging safe performance. They strengthen motivation and pride in adhering to safety values. When used properly, they complement—not replace—disciplinary or corrective systems.

**Q369. Answer: A**

**Explanation**: Simplifying complex information with relatable examples enhances comprehension and recall. Workers engage more effectively when content feels relevant and clear. Overly technical explanations reduce understanding and application on-site.

**Q370. Answer: C**

**Explanation**: Mentorship nurtures leadership capacity by transferring knowledge and modeling effective behaviors. It helps emerging leaders understand communication, risk management, and motivational techniques. Discipline or cost-saving goals are secondary outcomes, not the primary intent.

**Q371. Answer: B**

**Explanation**: The success of leadership initiatives is best measured by observable improvements in worker safety behavior and performance indicators. This includes reduced incident rates, increased reporting, and stronger participation in safety programs. Paperwork or meeting counts reflect activity, not genuine behavioral impact or cultural growth.

**Q372. Answer: A**

**Explanation**: Prioritizing safety sets a strong cultural precedent and demonstrates moral leadership. It reinforces that productivity should never compromise worker welfare. Effective communication about long-term benefits, such as reduced downtime and improved morale, aligns safety with operational goals.

**Q373. Answer: B**

**Explanation**: When leaders actively engage with workers during inspections, they promote transparency and trust. This presence encourages honest communication about safety issues. It transforms inspections from compliance exercises into collaborative learning experiences that strengthen organizational safety culture.

**Q374. Answer: B**

**Explanation**: Servant leaders focus on supporting their teams by providing resources, training, and removing obstacles to safety. This approach enhances trust and engagement. It contrasts with authoritarian leadership, which prioritizes control over collaboration, and often leads to compliance without commitment.

**Q375. Answer: A**

**Explanation**: Open reporting of hazards and near misses shows psychological safety and trust in leadership. It indicates that employees feel empowered and unafraid of retaliation. Restricted communication channels or limited feedback flow suggest a weak or fearful culture.

**Q376. Answer: D**

**Explanation**: Sharing lessons learned with the entire workforce ensures knowledge transfer and promotes continuous improvement. Corrective training reinforces changes and prevents recurrence. Concealment or limited circulation diminishes organizational learning and accountability.

**Q377. Answer: B**

**Explanation**: Inconsistency in leadership messages creates confusion, reduces credibility, and weakens adherence to safety expectations. Employees may perceive mixed priorities or lose motivation. Clear, consistent communication aligns understanding and reinforces shared safety objectives.

**Q378. Answer: A**

**Explanation**: Storytelling connects safety lessons to real experiences, making them relatable and memorable. Emotional engagement helps workers internalize messages and recall them during critical moments. Purely technical or humor-driven content lacks the impact needed for behavioral change.

**Q379. Answer: B**

**Explanation**: Peer observation programs empower employees to look out for one another and address risks collaboratively. They encourage mutual accountability and foster shared responsibility for safety. The goal is prevention and learning, not punishment or replacement of formal oversight.

**Q380. Answer: A**

**Explanation**: Immediate, specific feedback reinforces correct behavior and clarifies mistakes while maintaining respect. Supportive delivery ensures learning without shame. Delays or public reprimands can cause defensiveness and reduce the willingness to improve or communicate openly.

**Q381. Answer: C**

**Explanation**: Encouraging pride in safe work habits aligns safety with personal and team identity. Workers who view safety as integral to professionalism and pride maintain higher standards. Downplaying safety or focusing solely on output weakens its cultural significance.

**Q382. Answer: A**

**Explanation**: Adapting communication to the audience's background ensures clarity and comprehension. Frontline workers may need practical visuals, while managers require strategic overviews. Uniform content fails to address varying needs and comprehension levels, leading to misunderstanding.

**Q383. Answer: B**

**Explanation**: Empathy allows leaders to connect with workers' experiences, understand barriers to safety, and provide appropriate resources or support. It strengthens trust and cooperation. While compassion is key, empathy complements—not replaces—accountability and rule enforcement.

**Q384. Answer: A**

**Explanation**: Supervisor involvement models expected behavior and shows commitment to the organization's safety vision. Active participation increases meeting credibility and encourages employees to take safety discussions seriously. Passive or absent leadership undermines motivation and engagement.

**Q385. Answer: D**

**Explanation**: Evaluations gauge the effectiveness of training by assessing knowledge retention, behavioral application, and comprehension gaps. This information guides future improvements. Compliance records alone do not ensure that employees can apply what they've learned safely.

**Q386. Answer: A**

**Explanation**: Consistent use of PPE by leaders signals genuine commitment to safety and encourages compliance by example. Workers imitate visible behaviors of authority figures. Inconsistent actions or remote oversight weaken credibility and influence.

**Q387. Answer: A**

**Explanation**: Structured feedback sessions promote clarification, collaboration, and trust restoration between teams. They prevent recurring conflicts and establish open communication norms. Blame or avoidance worsens disconnection and hampers problem resolution.

**Q388. Answer: C**

**Explanation**: Translating materials and using inclusive examples ensure accessibility and respect for all cultural groups. Culturally responsive communication strengthens engagement and comprehension. Ignoring differences creates misunderstandings and may alienate part of the workforce.

**Q389. Answer: A**

**Explanation**: Safety climate reflects how employees perceive leadership's values, actions, and priorities regarding safety. A positive climate indicates trust, openness, and alignment between stated policies and actual behavior. It is a leading indicator of cultural maturity and engagement.

**Q390. Answer: A**

**Explanation**: Prompt and transparent communication prevents speculation and maintains trust. Providing accurate information from leadership reassures employees and supports emotional stability. Delayed or limited updates allow misinformation to spread, eroding confidence in management.

**Q391. Answer: B**

**Explanation**: Cross-functional committees bring diverse perspectives together, fostering creative solutions and shared accountability. They break down silos and ensure safety improvements are comprehensive. Restricting participation or using committees as paperwork exercises limits their impact.

**Q392. Answer: A**

**Explanation**: Regular refreshers and on-the-job assessments reinforce key principles and update workers on new standards. This ongoing learning approach maintains competency and awareness. One-time sessions or minimal reminders lead to knowledge decay and complacency.

**Q393. Answer: A**

**Explanation**: Supportive responses build psychological safety and encourage continued reporting. Leaders should recognize the value of near-miss reports as opportunities to prevent future incidents. Criticism or neglect discourages proactive communication and risk identification.

**Q394. Answer: D**

**Explanation**: Transformational leaders motivate workers by aligning safety with shared purpose and identity. They inspire through vision, trust, and empowerment. This intrinsic motivation fosters lasting commitment, unlike punitive or transactional approaches focused on compliance alone.

**Q395. Answer: A**

**Explanation**: Coaching aims to enhance understanding, confidence, and ownership, helping employees improve performance proactively. It emphasizes mentorship and skill-building, while supervision ensures compliance and task completion. Effective leaders balance both approaches to sustain safety excellence.

**Q396. Answer: A**

**Explanation**: Sustaining momentum requires consistent reinforcement through reviews, recognition, and celebration of progress. Acknowledging milestones keeps motivation high and aligns attention with long-term objectives. One-time announcements or frequent shifts in focus weaken commitment and continuity of effort.

**Q397. Answer: C**

**Explanation**: Engaging workers in identifying hazards and crafting corrective actions builds ownership and accountability. This participation ensures solutions are practical and fosters a culture of shared responsibility. When workers feel included, safety becomes a personal and collective commitment rather than a top-down mandate.

**Q398. Answer: A**

**Explanation**: Promptly addressing audit findings and including workers in corrective measures demonstrates transparency and responsiveness. This practice builds confidence in leadership's integrity and commitment to progress. Delayed or concealed responses reduce trust and impede learning opportunities.

**Q399. Answer: A**

**Explanation**: Credibility ensures that leadership influence is genuine and respected. When leaders act consistently with stated values, employees internalize those standards. Without credibility, even well-designed safety programs fail because trust is the foundation of cultural adoption and behavioral change.

**Q400. Answer: B**

**Explanation**: The ultimate goal is cultivating a sustainable safety culture where everyone takes ownership of safe behavior and continuous improvement. Leadership and communication foster engagement, while training builds competence. Compliance and cost control are secondary to developing enduring safety excellence through collective commitment.

Made in the USA
Monee, IL
25 June 2026

55801171R00063